Thematic Lessons and Materials for Teaching Spelling, Phonics, Science, Social Studies and More in Grades 1-3

Volume 1 of 2

70 Times 7 Math

Published by 70 Times 7 Math

THEMATIC LESSONS AND MATERIALS FOR TEACHING SPELLING, PHONICS, SCIENCE, SOCIAL STUDIES, AND MORE IN GRADES 1-3: VOLUME 1 OF 2

ISBN (Paperback Edition): 978-1-954796-69-0

Printed and bound in the United States of America

Published by 70 Times 7 Math

Visit www.habakkuk.net

TABLE OF CONTENTS

Volume 1 Book

Note: If you are unable to teach the Biblical themes in your school (i.e., "The Birth of Christ," "Musical Experiences and The Psalmist David," "Easter," "My Church," and "Joseph"), there are extra lessons at the end of Volume 1 that can be taught in their place.

Volume 2 Book

JANUARY

20. Martin Luther King, Jr. Day (3rd Monday of January)	long *e* /ē/	*ea, ee, eo, e, y*	
21. Growing Up Healthy (Community Helpers: Doctors and Dentists)	/b/	*b, bb*	

FEBRUARY

OPTION 1 FOR /D/ WORDS: 22. Groundhog Day (Feb. 2)	/d/	*d, de*	
OPTION 2 FOR /D/ WORDS: Friendship	/d/	*d, dd, de, -ed*	
23. Saint Valentine's Day	*r*-controlled vowels /ar/	*ar, ear, are, our*	
24. Presidents Day (3rd Monday in Feb.): George Washington's Birthday; Abraham Lincoln's Birthday	/v/	*v, ve, f*	
25. Continents and Cultures (A Trip to Japan; A Trip to Spain)	/p/ and the *sp* consonant blend	*p, pe, sp*	

MARCH

OPTION 1 FOR /Ī/ WORDS: 26. Weather in March (Air and Wind)	long *i* /ī/	*i*, *i* with final *e*, *y*, *ie*, *igh*	
OPTION 2 FOR /Ī/ WORDS: Self, I'm Special	long *i* /ī/	*i*, *i* with final *e*, *y*, *eye*, *eigh*	
OPTION 1 FOR /N/ WORDS: 27. Saint Patrick's Day (March 17)	/n/	*n*, *gn*	
OPTION 2 FOR /N/ WORDS: Safety Precautions (Community Helpers: Policemen)	/n/ and syllabic *n* in *suddenly*	*n*, *ne*, *gn*, *kn*, *en*	
28. Musical Experiences and The Psalmist David	/ŋ/	*ang*, *ing*, *ong*, *ung*	
29. The Spring Season	/g/ and /ŋ/	*g*, *gg*, *ng*	

APRIL

OPTION 1 FOR /S/ WORDS: 30. Easter (**Note:** If Easter is in March or later in April, feel free to switch a couple of lessons around.)	/s/	*s*, *ss*	
OPTION 2 FOR /S/ WORDS: Our Five Senses and Special People	/s/	*s*	
OPTION 3 FOR /S/ WORDS: The Four Seasons (Winter, Spring, Summer, Fall)	/s/	*s*, *ss*, *ce*, *c*	
31. Weather in April	long *a* /ā/	*a*, *a* with final *e*, *ai*, *ay*	
32. Earth Day	*r*-controlled vowels /ər/	*er*, *ir*, *ur*, *ear*, *or*	
OPTION 1 FOR /Z/ WORDS: 33. Caterpillars, Butterflies, and Other Insects	/z/	*zz*, *ze*, *se*, *s*, *es*	
OPTION 2 FOR /Z/ WORDS: Zoo and Rainforest Animals	/z/	*z*, *zz*, *s*	

MAY

OPTION 1 FOR /T/ WORDS: 34. Dinosaurs	/t/	*t, te*	
OPTION 2 FOR /T/ WORDS: Magnets	/t/	*t, tt, te*	
OPTION 3 FOR /T/ WORDS: Communication (Community Helper: Postman)	/t/	*t, te, tt*	
OPTION 4 FOR /T/ WORDS: Learning to Count (Money)	/t/ and the suffix *-ty*	*t, -ty*	
35. Families, Mother's Day, Father's Day	*r*-controlled vowels /ər/	*er, ir, ur, or*	
36. My Church	/ch/	*ch, tch*	
37. Joseph	/j/	*j, ge, g, dge*	

JUNE

38. Shapes	syllabic / and /əl/	*le, al, el*	
OPTION 1 FOR /L/ WORDS: 39. Colors and Colored Pigments	/l/ and /əl/	*l, ll, le*	
OPTION 2 FOR /L/ WORDS: Flag Day (June), Independence Day (July)	/l/	*l, le, ll*	
40. Water	/w/ and the *sw* consonant blend	*w, wh,* and the *sw* blend	
41. Feelings and Child Safety—Good and Bad Touches	/h/	*h, wh*	

Extra Lessons in Volume 1 for Repeated Themes and Repeated Sounds

Preface

Thematic Lessons and Materials for Teaching Spelling, Phonics, Science, Social Studies, and More in Grades 1-3 comes in two volumes that are sold separately. As suggested by the title, the books have four main objectives and many other secondary objectives. The **MAIN OBJECTIVES** are (1) for students to learn the spelling words, which relate both phonetically and thematically; (2) to learn different spelling patterns of each week's emphasized sound; and (3-4) to learn about the social studies or science theme in focus. (Most lessons list the specific social studies or science objectives for that lesson.) By planning lessons and choosing spelling words that correspond with a theme, students are provided ample opportunities to read and write the spelling words prior to the test while also learning about social studies or science.

The lessons come with a **spelling word search**, a set of **spelling word flashcards**, and worksheets and activities to help students learn about alphabetizing, contextual analysis, phonics, homophones, prefixes and suffixes, compound words, parts of speech, penmanship and writing skills. (See pages 14-20 for a more thorough description of these **SECONDARY OBJECTIVES** and suggestions on how to teach them.)

The **answer keys** are sold separately under the title *Thematic Lessons and Materials for Teaching Spelling, Phonics, Science, Social Studies, and More in Grades 1-3:* ***Answer Key for Volume 1*** and the corresponding *Thematic Lessons and Materials for Teaching Spelling, Phonics, Science, Social Studies, and More in Grades 1-3:* ***Answer Key for Volume 2***.

The table of contents suggests the lessons that should be covered each month because many of the lessons are seasonal and holiday related. A few of the lessons are also Biblically related. These include "The Birth of Christ," "Musical Experiences and The Psalmist David," "Easter," "My Church," and "Joseph." If you are unable to teach the Biblical themes in your school, there are extra lessons at the end of Volume 1 that can be taught in their place.

Together, Volumes 1 and 2, which are sold separately, provide a list of spelling words for nearly every sound in the English language, including the short vowel sounds, long vowel sounds, vowel digraphs, diphthongs, *r*-controlled vowels, consonant sounds, syllabic *l*, and all consonant digraphs except for the /zh/ sound. The chart with the table of contents lists each theme, the emphasized sound for the theme, and the spelling patterns that make the sound. For example, the long *e* sound can be studied during a theme on "Growing Up Healthy," and the spelling patterns taught are *ea* as in *peach*, *ee* as in *sleep*, *y* as in *dairy*, and *ey* as in *honey*.

There are also several optional **bonus spelling words** included with most lessons. The bonus words consist of compound words that include one of the regular spelling words (such as *beefsteak*), spelling words with inflectional endings added to them (e.g., *cleans*, *cleaning*, *cleaned*), and spelling words that are extra challenging, such as *protein*. It is suggested that you give students a sticker for their incentive charts for each bonus word spelled correctly and/or that you let a correctly spelled bonus word take the place of one of the regular spelling words that has been misspelled.

There are 19 weekly lessons available in Volume 1 and 22 lessons in Volume 2, but some of the lessons have alternative themes so that you have an option of which theme to teach for the emphasized sound. As an example, you can choose between a circus and a transportation theme when studying words with the /k/ sound. Volume 1 also has several alternative lessons included at the back of the book. Two math themes are also included in Volume 2.

At the end of most lessons, there is a **multiple-choice science or social studies test** with bubbles to fill in that can be used to assess individual learning and to verify that predetermined objectives have been met. To help prepare students for the tests, a set of **question cards** are available with the lessons to use for unit games, including the **"Thematic Units Game"** that is described on pages 28-29. The cards list what lesson they are from, and the answers can be found with that lesson under "Key Questions for Discussion" on the page after the list of spelling words. (One exception is that answers to unit questions in lesson 27 of Volume 2 are included on the "What Would You Do?" page.)

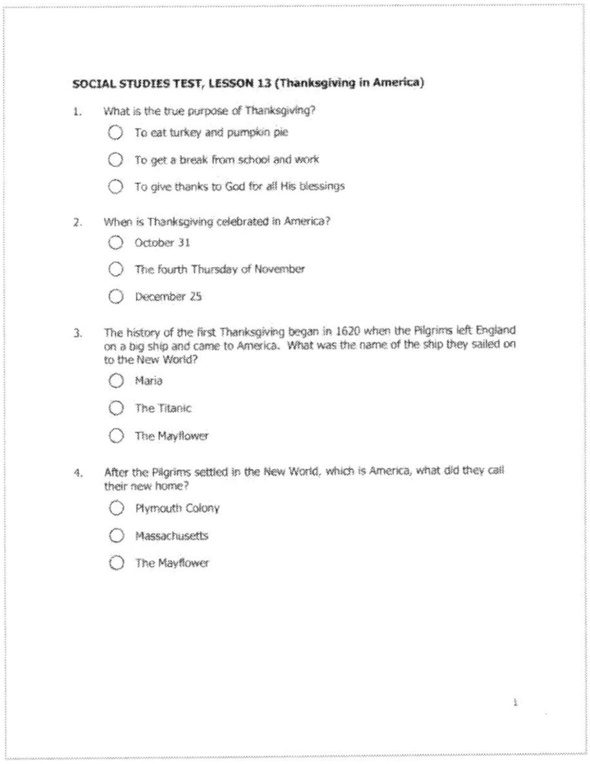
SOCIAL STUDIES TEST, LESSON 13 (Thanksgiving in America)

1. What is the true purpose of Thanksgiving?
   - To eat turkey and pumpkin pie
   - To get a break from school and work
   - To give thanks to God for all His blessings
2. When is Thanksgiving celebrated in America?
   - October 31
   - The fourth Thursday of November
   - December 25
3. The history of the first Thanksgiving began in 1620 when the Pilgrims left England on a big ship and came to America. What was the name of the ship they sailed on to the New World?
   - Maria
   - The Titanic
   - The Mayflower
4. After the Pilgrims settled in the New World, which is America, what did they call their new home?
   - Plymouth Colony
   - Massachusetts
   - The Mayflower

1

Computer-based social studies and science tests: The same multiple-choice questions that are included at the end of most lessons are also available online as computer-based tests. For free access to the tests, please visit the Habakkuk Educational Materials website at https://www.habakkuk.net/, click on "Free Computer-based Tests" from the homepage, and then click the image with an illustration of this book to be directed to the tests. (See page 13 for more information.)

Throughout the pages of this book, you will incorporate different teaching methodologies to meet individual learning needs, such as direct instruction and hands-on activities. You will find several **science experiments** as well as many **games** and other hands-on activities to help make learning an exciting experience, thus aiding their ability to comprehend the material. (If you are looking for a method for grading a student's participation and understanding of a given science experiment, consider the "Evaluation of Science Experiments" paper included on page 31.)

For some of the lessons, you will also find ideas for integrating **art**, **math**, **and other activities** into the themes. When studying Thanksgiving in Volume 1, for example, students can earn feathers (like Native Americans did) for different accomplishments throughout the week and then use them to make Indian headdresses. They also make a Thanksgiving booklet about the first Thanksgiving Day during this unit. During a theme on "Growing Up Healthy" in Volume 2, students will use paints and different vegetables to make vegetable prints. Other thematic activities suggested for this theme include making vegetable soup with the recipe provided, preparing a graph that depicts least- to greatest-liked vegetables of the class, and playing a categorization game to help students learn the five basic food groups.

Geography is even integrated into some lessons, particularly when it has to do with learning the 7 continents and 5 oceans, but world maps are also used to help students find some countries. When studying Thanksgiving, for instance, it is suggested that students find England on the map (where the Pilgrims set sail from in the Mayflower). Then they can locate the American state of Massachusetts to visualize where the Pilgrims arrived at in the New World.

Also note that on page 24 there is a **list of supplies** that will be used regularly throughout the year. When additional supplies besides these are needed, they are listed with the lesson. In addition, there is a list of ideas for **field trips** that are suggested for several lessons so that you might plan for these ahead of time, as well as a **suggested weekly schedule for teaching the spelling lessons**.

To contact Habakkuk Educational Materials or 70 Times 7 Math, please visit the website below.

https://www.habakkuk.net/

Access to Free Computer-Based Tests

Computer-based tests designed specifically to assess a student's knowledge of the information covered in this book are available by visiting the Habakkuk Educational Materials website. Follow the two steps below.

1. Go to https://www.habakkuk.net/. From the homepage, click on "Free Computer-based Tests." Then click on the image with an illustration of this book to be directed to an LMS.
2. From the LMS, you should see the title of the book you are using. Click the downward arrow beneath the title (it's on the right-hand side) to see the available tests. Then click on the test you want to take, followed by enter.

After submitting an answer to a question, the computer will notify test takers if their answer is correct or incorrect. After entering their answer to the final question and clicking "Continue," a percentage grade will be available and a "Review Course" option to review any incorrect answers will also be accessible.

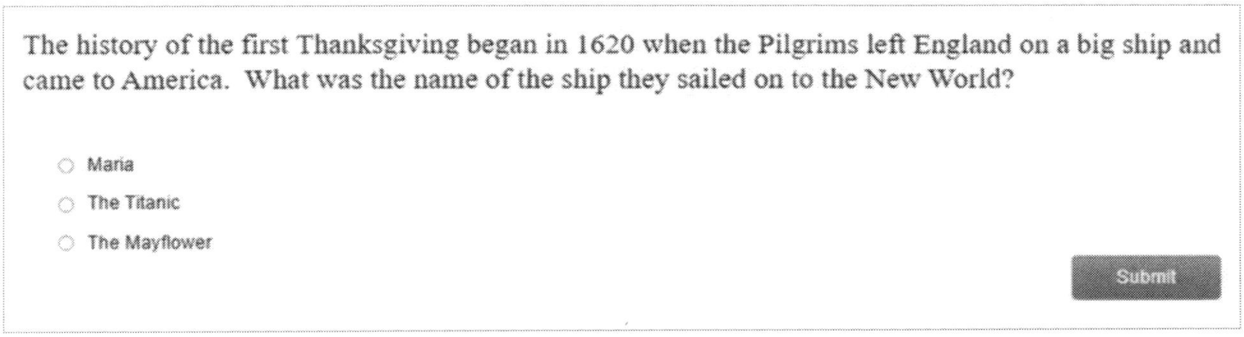

The tests serve two purposes:
(1) to assess a student's mastery of the content in this book; and
(2) to be used for review games.

Using the tests for review games: The computer-based tests can be used in correspondence with various board games that have a pathway from start to finish. Students playing the game would answer one of the questions (most of which are multiple choice), and if the computer confirms that the answer is correct, the student could roll dice or spin a spinner and move his or her playing piece the corresponding number of spaces on the path. If you visit the "Free Teaching Materials" page of the Habakkuk Educational Materials website, there are free gameboards that can be printed on cardstock. (Click on "Board Games and More" to access the games.)

Instructions for Teaching the Secondary Objectives

Papers to use for direct instruction purposes: You will find the worksheets and activities listed below with the lessons included in this book. When teaching students about alphabetizing, compound words, parts of speech, and so forth, you will need a matching worksheet to position under a Smart Document Camera or a similar device. If you only have the book with the answer key, you may make a copy of the page from a student's book for non-commercial, instructional purposes or put a sheet of transparency paper over it. Once students are familiar with the skill, you can wait until they have finished an assignment and turned it in (or closed their workbooks) before completing the worksheet on your classroom screen.

- **Alphabetizing words:** On the right half of each week's list of spelling words, there is a place for alphabetizing the spelling words.

  Teaching students how to alphabetize words: Each week, position the list of spelling words under a Smart Document Camera or a similar device. Sing the "Alphabet Song" slowly, pausing after each letter to search the list for a word that begins with the letter. When you find one, cross out the word and write it beside number one. When more than one word begins with the same letter, put a checkmark by them and underline the second letters before singing the song again to determine which letter comes first. When the exercise is complete, students alphabetize the words on worksheets of their own.

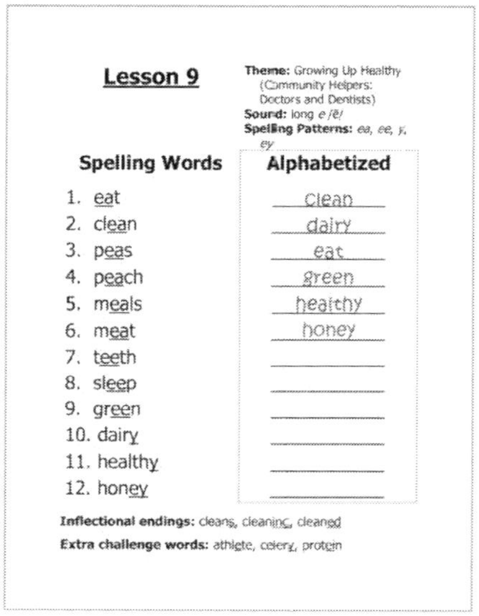

Lesson 9

Theme: Growing Up Healthy (Community Helpers: Doctors and Dentists)
Sound: long *e* /ē/
Spelling Patterns: *ea, ee, y, ey*

Spelling Words	Alphabetized
1. eat	clean
2. clean	dairy
3. peas	eat
4. peach	green
5. meals	healthy
6. meat	honey
7. teeth	
8. sleep	
9. green	
10. dairy	
11. healthy	
12. honey	

Inflectional endings: cleans, cleaning, cleaned

Extra challenge words: athlete, celery, protein

- **Context Clues (Contextual Analysis):** To acquaint students with the contextual analysis method for identifying unknown words when reading while also teaching them facts about the current week's social studies or science theme, each week's lesson includes a worksheet containing thematic sentences with blank spaces provided to insert missing spelling words. A student reads the sentences and fills in each blank with a spelling word that makes sense in the context of the sentence. For example, "*Beef and other meats belong in the protein group.*" Because spelling words will relate phonetically, the student can afterwards circle the letters, *ee* in the example provided, that make the emphasized sound. (Note that the sentences on these worksheets are also meant to be used by the teacher during spelling tests when she reads a word, uses it in a sentence, and then repeats the word.)

- **"Spelling Patterns of Sounds Heard in the English Language" chart:** A description of this chart is included on an upcoming page.

- **Spelling Words Riddle Game:** If the spelling words have several spelling patterns of the emphasized sound, a "Spelling Word Riddles" game has been provided with the lesson to help students learn the different spelling patterns. During a theme on "Growing Up Healthy," for example, all spelling words have the long *e* sound, /ē/, and can be spelled as ***ea*** as in *clean*, ***ee*** as in *sleep*, ***y*** as in *baby*, or ***ey*** as in *honey*. The long *e* sound is the same as the name of the letter *e*.

  Teachers reproduce the letter strips on light blue or yellow cardstock as instructed, cut out the pieces, and then attach a magnetic strip to the back of each piece. The yellow pieces with the key sound are lined up horizontally towards the top of a magnetic board. The light blue strips are positioned vertically below them. The teacher reads a riddle and calls on a student to identify the correct spelling word. For instance, one of the riddles is, *"In this word, the long e sound is spelled e-a. This is what your hands should be before you eat."* After listening to the riddle, a student would identify the word *clean* and then move the yellow magnetic strip with the letters ***ea*** between the light blue letters ***cl*** and ***n*** to form the word ***clean***.

  This game illustrates how the spelling words relate phonetically and thematically, and although each student should be encouraged to study the spelling words at home, activities such as this are provided to help students recognize the spelling patterns of each week's emphasized sound.

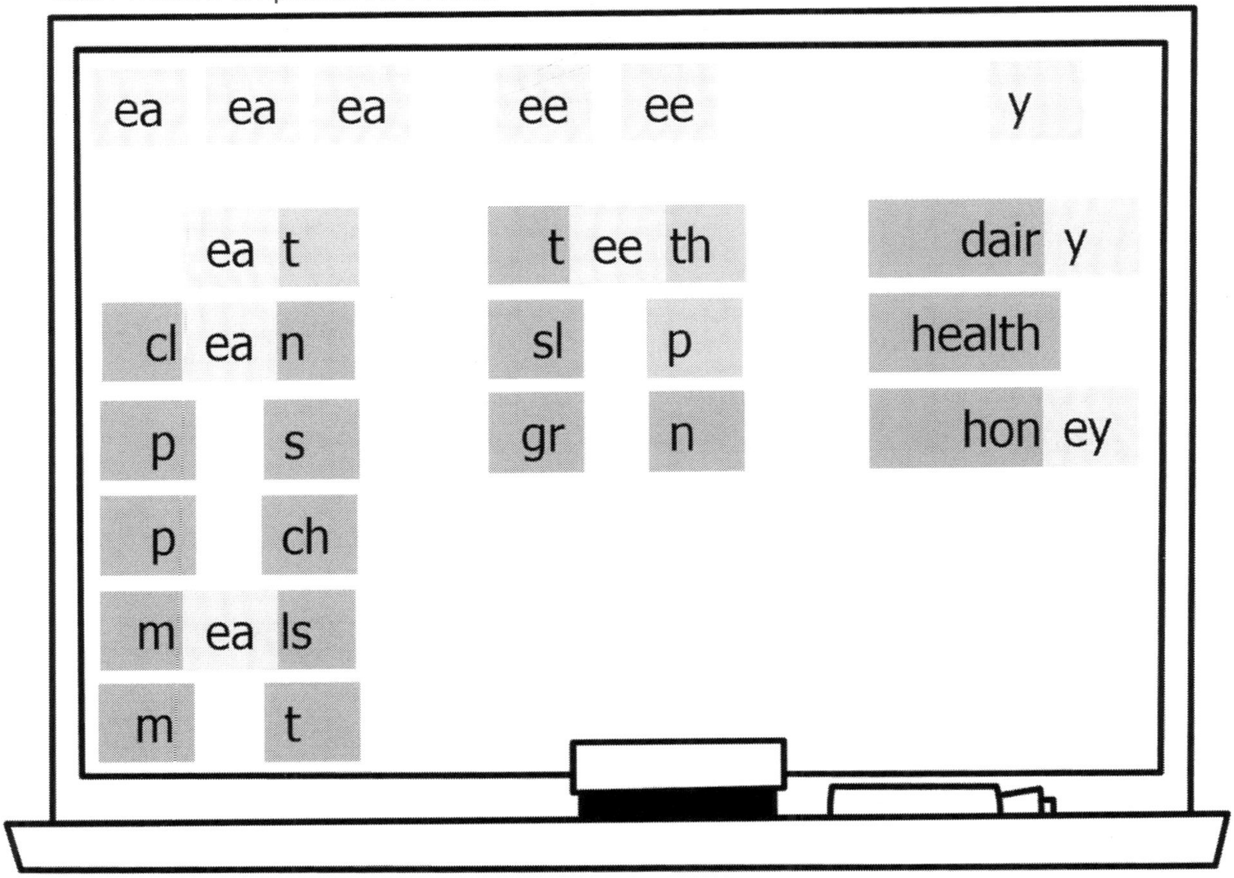

- **Spelling Patterns of Sounds Heard in Our Spelling Words:** Most lessons also include a worksheet titled "Spelling Patterns of Sounds Heard in Our Spelling Words." The spelling words are listed, but the key sound is omitted. Students fill in the missing letter or letters that make, for example, the long *e* sound in the word and use a different colored pencil for the different spelling patterns that make the sound. In the example worksheet shown, students would use a blue colored pencil for the spelling pattern *ea*, a red colored pencil for the spelling pattern *ee*, a green colored pencil for the spelling pattern *y*, and a purple colored pencil for the spelling pattern *ey*. At the bottom of the page, they categorize the words by their spelling patterns.

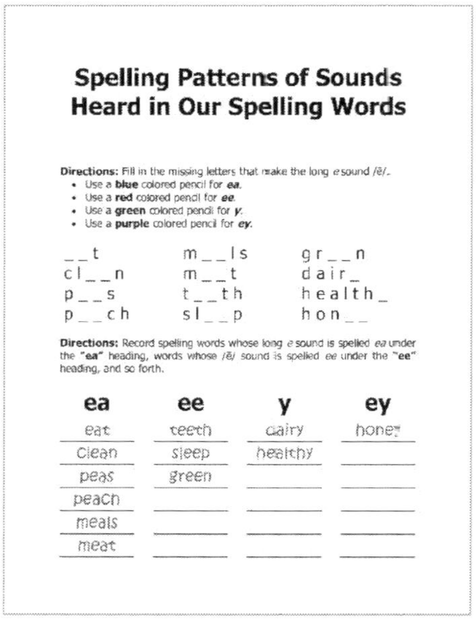

- **Spelling Word Search:** Each lesson also includes a word search where students can look for the hidden spelling words vertically and horizontally. The word search is meant for younger students (grades first through third), and thus none of the words are recorded backwards or diagonally.

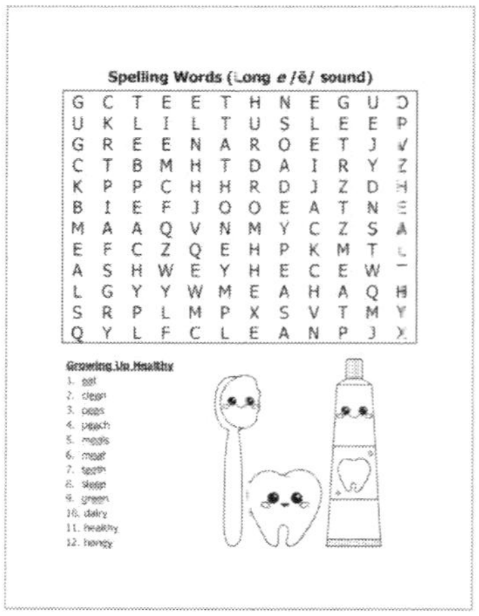

- **Homophones:** Homophone words sound alike, but they are spelled differently and have different meanings, such as *meat* and *meet*. If a spelling word has a homophone, a worksheet having the same format as that shown is included with the lesson. Through the use of illustrations that can be colored as well as sentences where the key words are used in context, students will learn to differentiate between homophones like *meat* and *meat* and *son and sun*.

- **Prefix / Root Word / Suffix:** Several of the lessons include a "Prefix / Root Word / Suffix" Worksheet. Essentially, these worksheets teach students what a prefix, a root word, and a suffix is; allows them to practice adding prefixes to the beginning of root words and suffixes at the end; helps them to learn one or more of the spelling words (any root words are taken from the week's spelling word list); and helps them to understand how the prefix and/or suffix changes the meaning of the root word. Moreover, the worksheets also help students to learn more about the science or social studies theme in focus, as the sentences where the words are used in context are theme related. In the example shown, students would determine what form of the spelling word *thank* (*thankful*, *unthankful*, or *thankfulness*) makes sense in the context of the sentences. For direct instruction purposes, complete the worksheet on your classroom screen after all papers have been collected or students have closed their workbooks. After the prefixes and suffixes have been inserted in the spaces preceding and following the root words, let students read the sentences towards the bottom of the page in unison.

Directions: Add the prefix and suffixes listed below to the root word *thank* to form a word that means "full of thanks," "not thankful," and "being full of thanks."

Prefix: *un-*
Suffixes: *-ful*, *-ness*

Prefix	Root Word	Suffix	
	thank	ful	
un	thank	ful	
	thank	ful	ness

Which word means "full of thanks"? thankful
Which word means "not thankful"? unthankful
Which word is a noun meaning "being full of thanks"? thankfulness

Directions: Insert a prefix or suffix in the space provided to form a new word that makes sense in the context of the sentence.

Thanksgiving is a special time to **thank** God and to remind us to be **thank**ful for the good things He provides. We must never be un**thank**ful, so every day, not just on Thanksgiving, we should show are **thank**fulness.

- A **prefix** is added to the beginning of a root word to form a new word (altogether). A **suffix** is added to the end of a root word to form a new word (thankful). Both prefixes and suffixes are **affixes**.
- **Derivative:** A root word with something added to it (e.g., a prefix or a suffix)
- **Rule:** When a consonant suffix is added to a word, just put the word and the suffix together. (Thankful)

- **Compound Words:** Many of the lessons include one or more "Compound Words" worksheets. (When two or more root words are used together to form one word, such as <u>farmyard</u>, it is called a compound word.) These worksheets teach students what a compound word is, helps them to understand that some compound words are closed (as in *toothbrush*) and some are open (as in *green beans*), allows them to practice forming compound words, and helps them to identify their use in context. They also help students to learn some of their spelling words (as at least one of the words used to form each compound is from their list of spelling words) and to learn more about the science or social studies theme in focus, as the sentences where the words are used in context are theme related. Besides the worksheets, many of the spelling word lists include optional bonus compound words, where a compound is formed from one of the regular spelling words.

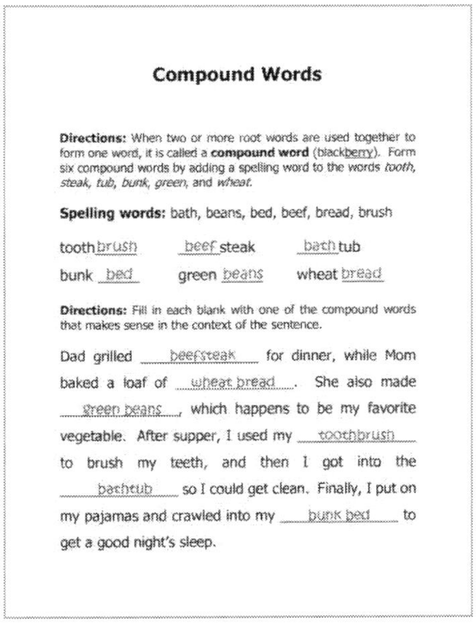

Compound Words

Directions: When two or more root words are used together to form one word, it is called a **compound word** (black<u>berry</u>). Form six compound words by adding a spelling word to the words *tooth*, *steak*, *tub*, *bunk*, *green*, and *wheat*.

Spelling words: bath, beans, bed, beef, bread, brush

toothbrush beefsteak bathtub
bunk bed green beans wheat bread

Directions: Fill in each blank with one of the compound words that makes sense in the context of the sentence.

Dad grilled beefsteak for dinner, while Mom baked a loaf of wheat bread. She also made green beans, which happens to be my favorite vegetable. After supper, I used my toothbrush to brush my teeth, and then I got into the bathtub so I could get clean. Finally, I put on my pajamas and crawled into my bunk bed to get a good night's sleep.

- **WEEKLY ORAL LANGUAGE:** Every lesson includes a "Weekly Oral Language" worksheet for students to complete. These include both asking and telling sentences with capitalization, punctuation, and sometimes spelling errors for the child to correct. The sentences relate to the week's theme, and any spelling errors were taken from the list of spelling words since they also relate to the theme. As can be seen from the illustration, each child will also be expected to identify the subject and verb of the telling sentence (as well as the helping verb whenever one is included) and to write his own sentence corresponding to the weekly theme. *Teaching Children about Capitalization, Punctuation, Apostrophe s, and Contractions (Language Book 2)* by Habakkuk Educational Materials can be purchased separately to give students the knowledge they need to complete this worksheet. It is also recommended that a teacher correct the worksheet on the classroom screen after all papers have been collected or students have closed their workbooks.

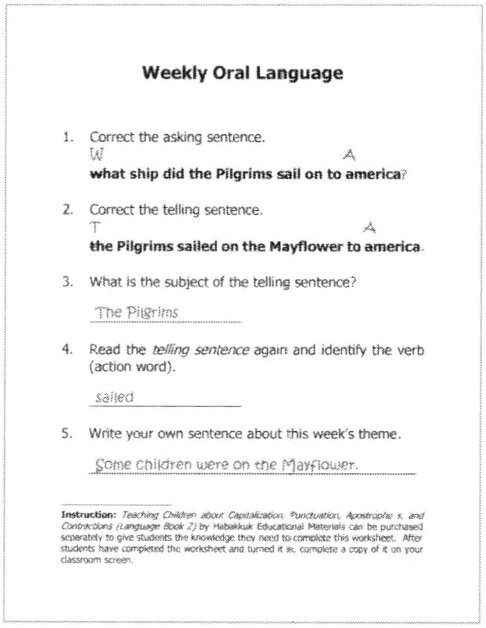

Weekly Oral Language

1. Correct the asking sentence.
   W A
   what ship did the Pilgrims sail on to america?
2. Correct the telling sentence.
   T A
   the Pilgrims sailed on the Mayflower to america.
3. What is the subject of the telling sentence?
   The Pilgrims
4. Read the *telling sentence* again and identify the verb (action word).
   sailed
5. Write your own sentence about this week's theme.
   Some children were on the Mayflower.

Instruction: *Teaching Children about Capitalization, Punctuation, Apostrophe s, and Contractions (Language Book 2)* by Habakkuk Educational Materials can be purchased seperately to give students the knowledge they need to complete this worksheet. After students have completed the worksheet and turned it in, complete a copy of it on your classroom screen.

To find the subject of the sentence
***A** kangaroo can jump 6 feet high,* you would ask:
"What can jump 6 feet high?" (*A kangaroo*)
Then find the verb of the sentence by asking:
"What can a kangaroo do?" (*jump*)
The helping verb *can* is in front of the verb *jump*.

- **Parts of Speech:** Each lesson includes a worksheet where students will identify the parts of speech and the subject (naming part) and predicate (action part) of a thematic sentence. *Teaching Children about Parts of Speech (Language Book 3)* by Habakkuk Educational Materials can be purchased separately to help students quickly learn the parts of speech. For further instruction, it is also recommended that a teacher correct the worksheet on the classroom screen after all papers have been collected or students have closed their workbooks. Note that Schoolhouse Rock videos available on YouTube can also help teach students about adjectives, interjections, and so forth.

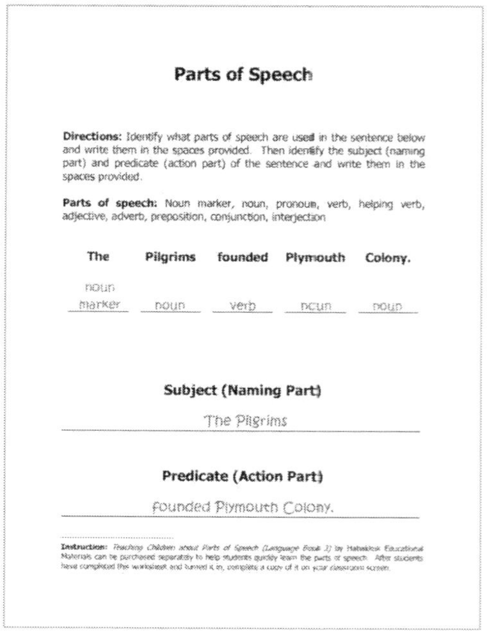

Parts of Speech

Directions: Identify what parts of speech are used in the sentence below and write them in the spaces provided. Then identify the subject (naming part) and predicate (action part) of the sentence and write them in the spaces provided.

Parts of speech: Noun marker, noun, pronoun, verb, helping verb, adjective, adverb, preposition, conjunction, interjection

The	Pilgrims	founded	Plymouth	Colony.
noun marker	noun	verb	noun	noun

Subject (Naming Part)

The Pilgrims

Predicate (Action Part)

founded Plymouth Colony.

Instruction: *Teaching Children about Parts of Speech (Language Book 3)* by Habakkuk Educational Materials can be purchased separately to help students quickly learn the parts of speech. After students have completed the worksheet and turned it in, complete a copy of it on your classroom screen.

- **Writing/Handwriting assignment:** Most lessons also include a writing and handwriting assignment relating to the week's theme. If one is not provided, you might want to use one of the thematic sentences from the "Context Clues" worksheet for the weekly handwriting lesson. Another option would be to ask students what they have learned about the theme in focus and to use one or more of the sentences they suggest as the handwriting assignment. You may make a copy of the reproducible writing paper on page 30 to use for the lessons.

  During the Thanksgiving unit, the teacher would position the handwriting assignment shown here of a family enjoying a Thanksgiving meal together under a Smart Document Camera or a similar device. The teacher would then discuss the correct stroke for each letter as he or she copies the incomplete sentence "I am thankful for..." onto the paper. Students would follow along with the teacher, copying the letters on papers of their own. (This will help them to remain on task and to finish the handwriting assignment in unison.) They can then finish the sentence on their own and add additional sentences. Note that you may want to give students the option of sharing their weekly writing papers in large group.

Handwriting directions: Color the picture. Then copy "I am thankful for" on the lines at the bottom of the page. Complete the sentence by telling what you are thankful for.

I am thankful for...

- **Addition and Subtraction Facts:** These worksheets include images relating to the week's theme and step-by-step directions for students to follow when solving the addition and subtraction problems.

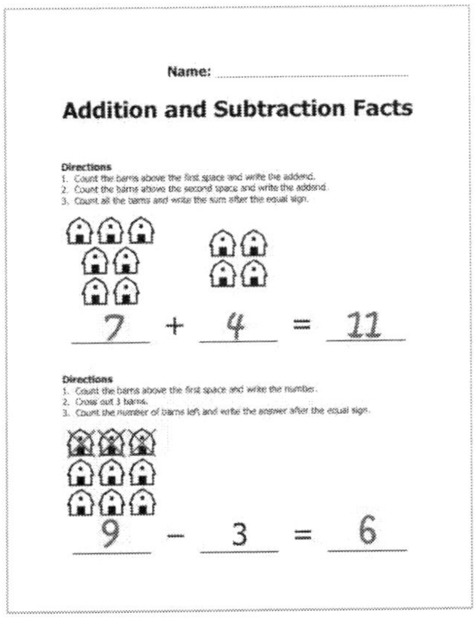

Name:

Addition and Subtraction Facts

Directions
1. Count the barns above the first space and write the addend.
2. Count the barns above the second space and write the addend.
3. Count all the barns and write the sum after the equal sign.

7 + 4 = 11

Directions
1. Count the barns above the first space and write the number.
2. Cross out 3 barns.
3. Count the number of barns left and write the answer after the equal sign.

9 – 3 = 6

- **Addition and Subtraction Word Problems:** Students color and cut out the pictures included with the worksheet. They record each equation as presented in the word problems and then use the cutouts to illustrate the problems and to find the answers.

  The worksheets can also be used by the teacher for direct instruction purposes on a magnetic board or with an overhead projector. If you are using them with an overhead projector, simply reproduce the page on transparency paper and cut out the images. To use them with a magnetic board, see the instructions below.

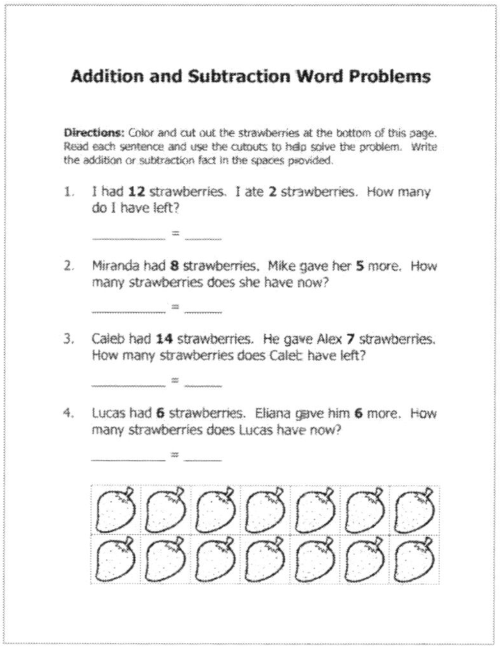

Addition and Subtraction Word Problems

Directions: Color and cut out the strawberries at the bottom of this page. Read each sentence and use the cutouts to help solve the problem. Write the addition or subtraction fact in the spaces provided.

1. I had **12** strawberries. I ate **2** strawberries. How many do I have left?
   ________ = ____
2. Miranda had **8** strawberries. Mike gave her **5** more. How many strawberries does she have now?
   ________ = ____
3. Caleb had **14** strawberries. He gave Alex **7** strawberries. How many strawberries does Caleb have left?
   ________ = ____
4. Lucas had **6** strawberries. Eliana gave him **6** more. How many strawberries does Lucas have now?
   ________ = ____

Using the images for direct instruction purposes on a magnetic board or for use in a math center: Reproduce the images on cardstock. You may want to color and laminate the pictures before cutting them out. Attach a magnetic strip to the back of each piece. Provide magnetic numerals and operation signs that include plus, minus, and equal signs.

Example problem: I had **12** strawberries. I ate **2** strawberries. How many strawberries do I have left?

1. Put twelve strawberries and the number 12 on the magnetic board. (Count the strawberries.)
2. Put the minus sign and the number 2 on the magnetic board. (Take away two strawberries.)
3. Put the equal sign and number 10 on the magnetic board. (Count the ten strawberries left.)

Spelling Word Games

Reproducible word cards are provided with each list of spelling words. It is suggested that you reproduce the pages on card stock so that they are not see through. You will need two copies of each set so that each card has a match for the matching game. Ideas for how the cards might be used in your reading or spelling centers or during small or large groups are provided below.

Matching Spelling Words Card Game: Besides using the matching game for your reading or spelling centers or during small groups, they can also be used during large group. Line up one set of the cards on one half of the chalkboard tray and the other set on the other half with the blank side of cards facing the students. A child could turn over a card on both sides of the tray to see if they make a match.

Spelling Words Bingo: The spelling word cards can also be drawn during a bingo game. Reproducible bingo cards are provided in this section. Students are instructed to fill in the grids randomly with one of the spelling words. You might have them fill in the grids while completing the week's "Context Clues" worksheet. After filling in a blank with a spelling word that makes sense in the context of the sentence, they could record the same word randomly on their grids.

Go Fish: Deal five cards to each player and place the others in a stack. (You will need to combine the spelling word cards with words from previous weeks so that you have enough cards to play with.)

Player 1: "Do you have the word _____?" (Example: If Player 1 has the word *peach* in her set, she might ask another player if he has that same word. Player 1 would show the word to the other player while identifying its spelling—"Do you have the word *peach*: *P-e-a-c-h, peach*?")

> <u>Player 2 has the card:</u> Player 2 sets *peach* down and spells it out. (Player 1 keeps the pair and has a second turn.)
>
> <u>Player 2 does not have the card:</u> Player 2 says, "Go Fish!" (Player 1 draws a card, and it is player 2's turn.)

Around the World: To use the cards for "Around the World," two students from the class would stand together, and the teacher would hold up one of the spelling words. The first child to read the word would move to the next child in line for a second race, and the game would continue in this manner.

Using the cards for a relay game: Students line up in teams at the chalkboard for a relay game. The teacher would read a spelling word card without showing it to the students, and the first team player to spell it on the board would have a tally mark recorded under his/her group's name. Count the tally marks at the end of the game to determine the winner.

The Wheel of Fortune Game (sort of!): Think of a sentence that includes spelling words from the current week and that relates to science or social studies information studied during thematic lessons. Draw spaces on the chalkboard to indicate the number of words in the sentence and the number of letters that make up each word. If you were using the sentence *Dairy is one of the five basic food groups*, your spaces would resemble those shown below. Students take turns guessing letters until a child can identify the whole sentence.

D _ _ _ _   _ _   o _ _   o _   _ _ _   _ _ _ _   _ _ _ _ _   _ o o d   _ _ o _ _ _.

Variation: Rather than using a sentence, draw spaces on the chalkboard to indicate the number of letters in a spelling word only and let students take turns guessing letters until they can identify the word.

Spelling Words Bingo		

Spelling Words Bingo		

Directions: Students randomly fill in the grids with one of the spelling words.

Materials Needed and Field Trip Suggestions

Materials Needed for Lessons

If materials other than what is included in this book is needed for a lesson, these have been listed on the first page after the list of spelling words. It is suggested that the materials listed below be available at the beginning of the school year, as they will be used regularly throughout the books.

- Students will need a notebook for their writing journals.
- White cardstock (Needed to prepare the spelling word cards included at the end of each lesson)
- Light blue and yellow cardstock (Used to prepare the cards used with the "Spelling Word Riddles")
- Magnetic strips
- Several shoeboxes with lids (Needed for several phonics games used in the book)
- Lunch bags
- Pencils, erasers, crayons, colored pencils, scissors, glue stick
- Magnetic numerals and operation signs that include plus, minus, and equal signs (optional)
- A sweatshirt with a kangaroo pocket or a similar item of clothing is needed for lesson 19 in Volume 1.

Field Trips Suggestions

<u>Volume 1</u>

- Fire Safety (Lesson 7): Visit a local fire station.
- Farm Life (Lessons 12-13): Visit a local farm.
- The Birth of Christ (Lesson 17): Deliver Christmas cards to a local nursing home or children's hospital. Plan to sing Christmas carols too!

<u>Volume 2</u>

- Saint Valentine's Day (Lesson 23): Plan a field trip to a local post office and mail valentine's cards to nursing home residents, to patients in a children's hospital, or to troops.
- Zoo and Rainforest Animals (Lesson 33): Field trip to the zoo

Thematic Spelling Lesson Schedule

Note: If you are teaching beginning readers who are not yet familiar with the alphabet and consonant blends sounds, you might want to wait until about the 4th week of school to start assigning spelling words. The alphabet sounds can usually be learned within a couple of months or less with Habakkuk Educational Materials' ***Digital Alphabet Flashcards***, and consonant blends and digraphs are amongst the first flashcards to be introduced with ***Phonics Book for Kids and Digital Flashcards for Teachers***. Please visit https://www.habakkuk.net/ to purchase these supplementary resources.

Monday

- Pass out the spelling words and read them aloud as a class.
- Spelling word search

Tuesday

- "Context Clues" worksheet
- "Spelling Patterns of Sounds Heard in Our Spelling Words" worksheet (if provided)

Wednesday

- Alphabetize the spelling words
- Spelling pretest (Circle *Pretest* on the paper used for spelling pretests and tests. Also, see the note below.)

Thursday

- Students write words misspelled on the pretest ten times each. Students who score 100% on the pretest can have free time during the spelling test.
- Spelling Word Riddles game (if one is provided)

Friday

- Spelling test (Circle *Test* on the paper used for spelling pretests and tests.)

Schedule the following worksheets into your lesson plan as well if one is provided: Homophones; Prefix / Root Word / Suffix; Compound Words; Weekly Oral Language; Parts of Speech; thematic writing/handwriting assignment; other thematic activities provided with the lesson; Thematic Units Game

Note: It is recommended that you give a **spelling pretest** each Wednesday to help individual students know which words to concentrate on. Unless a child scores 100% on this midweek exam, his pretest can be sent home with him, and he can be retested on Friday. During **spelling tests**, words are read, used in a thematic sentence (from the "Context Clues" worksheet provided with the lesson), and then repeated. To avoid disruptions during the test, instruct students to raise their hands if they need additional time to write a word before a second word is given. Also, to help students keep their eyes on their own papers, have them stand two pocket folders up on their desks in such a way as to create a barrier around their papers as in voting booths. Remind students to set up their "**voting booths**" before tests and pretests.

Spelling Patterns of Sounds Heard in the English Language

Assemble the chart: For access to the chart materials, please visit the "Free Teaching Materials" page of the business website at https://www.habakkuk.net/ and click on the illustration of the "Spelling Patterns of Sounds Heard in the English Language" chart. It is recommended that you use white cardstock to print the available title, sound cards, and word cards for the chart. Although not evident by the chart illustration on the following page, the sound cards are color coded, and the spelling patterns that make the key sound are also color coded differently on the word cards and not underlined as shown. (If you prefer, you can use a die cutting machine to cut out letters from laminated construction paper to spell the title.) Tape the title "Spelling Patterns of Sounds Heard in the English Language" across your classroom wall. Tape colored ribbon that is about 1½-inches wide and a couple of feet long to the back of each sound card. Tape the colored sound cards (not the plain sound cards) below the title or use a clothespin to clip them to a clothesline you have hung to the wall beneath the title. (The recommended order of the sound cards from left to right is listed below.) Use paperclips to fasten the word cards with different spelling patterns of the sound to the ribbon dangling from the sound card. It is best to assemble the entire chart at the beginning of the year. One of the reasons for this is so that you know how much wall space will be needed. The word cards clipped to the sound card ribbons can be facing backwards until the week you introduce the sound during spelling lessons. When studying "Growing Up Healthy," for example, you will need the /ər/ sound card and the words *moth**er***, *g**ir**l*, *n**ur**se*, ***ear**th*, *pol**ar***, and *hon**or***, as these cards illustrate different spelling patterns of the /ər/ sound. Again, a different color has been used for each spelling pattern. (Note that you might need to turn the cards back over during tests and pretests.)

Order of sound cards from left to right (begin with /ă/ and end with /ng/)
Short *a* /ă/, Short *e* /ĕ/, Short *i* /ĭ/, Short *o* /ŏ/, Short *u* /ŭ/, Long *a* /ā/, Long *e* /ē/, Long *i* /ī/, Long *o* /ō/, /ü/ as in *moon*, /u/ as in *look*, /ou/, /oi/, /ar/, /or/, /er/, /ĕər/ as in *bear*, /b/, /d/, /f/, /g/, /h/, /j/, /k/, /l/, Syllabic *l*, /m/, /n/, Syllabic *n*, /p/, /s/, /t/, /v/, /w/, /y/, /z/, /ch/, /sh/, /th/ in *thank* and /th/ in *father*, /ng/

How to use the chart: Students should have regular opportunities to review the sounds and spelling patterns of the chart. They can name the colored letters on a word card, identify the sound, and then read the word. As an example for the /or/ sound, students would recite "*o-r* /or/ *corn*, *o-r-e* /or/ *store*, o-o-r /or/ *floor*" before moving to a different sound on the chart. Another option would be to have students identify the sound of the colored cards and then read the words hanging below them.

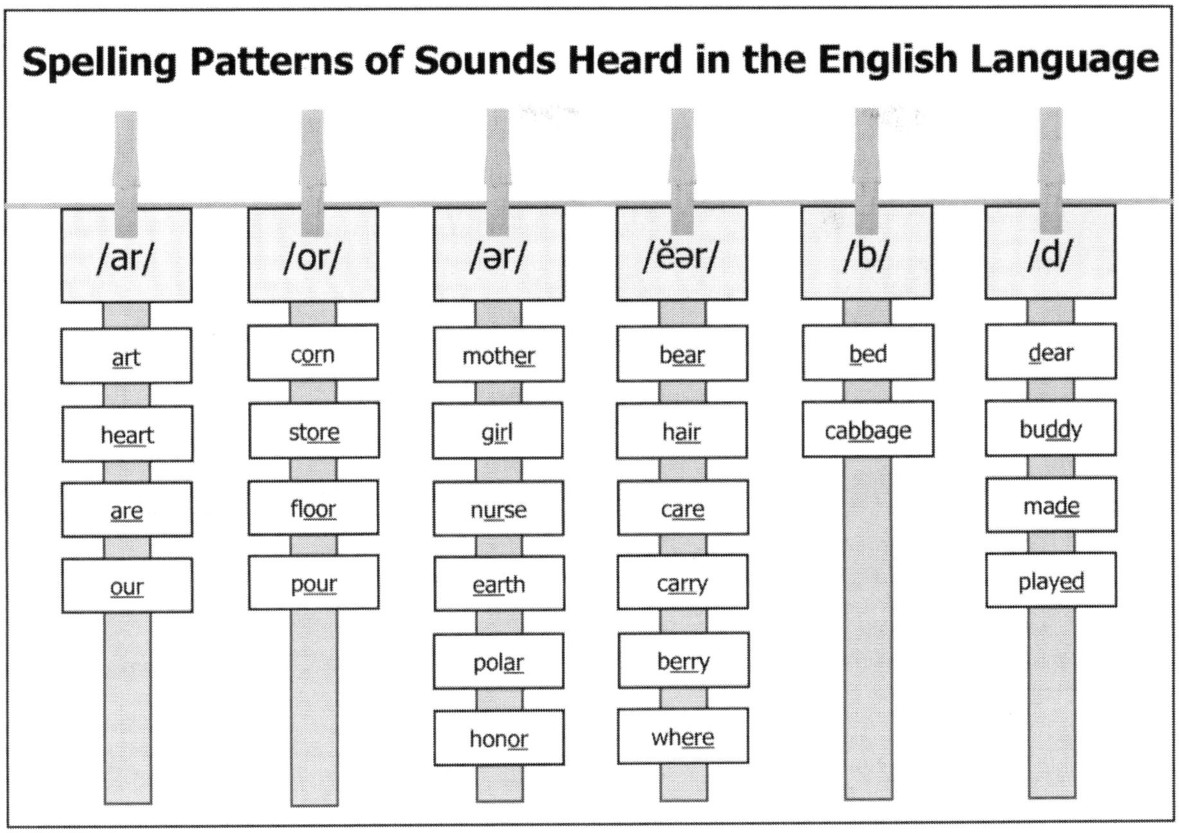

Using the chart cards for a matching game: The plain sound cards (those that are not colored) are used ONLY for the matching games. Randomly line up the ones you are using (such as /ar/, /or/, and /ər/) on one end of the chalkboard tray with the blank side of cards facing the students. Remove one word card from each ribbon of the chart to represent the sound cards you are using and line them up on the other end of the chalkboard tray. Examples of different matching games are listed below.

- */ē/*, *dair**y*** (These cards would be considered a match because the colored letter *y* makes the long *e* sound.)
- ***art*** and ***heart*** or ***corn*** and ***floor*** (*Do these r-controlled vowels have the same sound?* ***Art*** and ***heart*** or ***corn*** and ***floor*** would be considered a match because the colored letters make the same sound.) (**Note:** For this matching game, you would remove two word cards from each ribbon you want to use for the game.)
- /ch/ and ***church***, /sh/ and ***chef***, /k/ and ***chorus*** (The /ch/ and ***church*** cards would be considered a match because the *ch* in *church* makes the /ch/ sound and not the /sh/ sound it makes in *chef* or the /k/ sound it makes in *chorus*.)

A child will choose a card from each end of the chalkboard tray in an attempt to match a word such as *dair**y*** with the sound made by the colored letter or with another word whose colored letter or letters make the same sound. If a match is made, the child would say something like, "The *y* in dairy makes the long *e* sound," or "the colored letters in both words make the /ē/ sound," while removing the cards from the tray. If a match is not made, the cards would be returned to their original positions.

Thematic Units Game

THEMATIC UNITS GAME: You will need to wait until the 8th lesson of Volume 1 before using this game so that cards are available for all three categories of the chart. However, you can begin using the unit questions as early as the 4th week of school (the 4th lesson of Volume 1) with other games.

The question cards can be reproduced on cardstock and cut out. Each week, after you have used the unit question cards for other available games provided with the lessons or those available on the "Free Teaching Materials" page of the Habakkuk Educational Materials website, you can add them to the Thematic Units Game bags for year-round reviews. (Instructions on how to prepare the game are included on the following page.) Question cards labeled as "Science Themes" are added to the "Seasons and Other Science Themes" bag; question cards labeled as "Social Studies" are added to the "Holidays and Other Social Studies Themes" bag; and question cards labeled as "Child Safety" are added to the corresponding bag.

If you play the game weekly throughout the school year, it will help students to retain the facts they learned from previous lessons. The cards list what lesson they are from, and the answers can be found with that lesson under "Key Questions for Discussion" on the page after the list of spelling words. One exception is that answers to unit questions in lesson 27 of Volume 2 are included on the "What Would You Do?" page.

This game is also meant to help students review for the multiple-choice social studies or science tests that are included with most lessons to assess individual learning and to verify that predetermined objectives have been met.

GAME VARIATIONS: You could use the social studies and science unit questions with virtually any gameboard that has a pathway from start to finish. A student draws a unit question and answers it. If he answers it correctly, let him spin a spinner or roll dice and then move the corresponding number of spaces down the path. Such games could also be used in science and social studies centers. Note that there are also **computer-based tests** that can be used with gameboards in place of the question cards. The tests are available by visiting the "Free Computer-based Tests" page of the Habakkuk Educational Materials website. Students playing the game would answer one of the questions (most of which are multiple choice), and if the computer confirms that the answer is correct, the student could roll dice or spin a spinner and move his or her playing piece the corresponding number of spaces on the path.

PREPARING THE THEMATIC UNITS GAME CHART: For access to the game materials, please visit the "Free Teaching Materials" page of the Habakkuk Educational Materials website at https://www.habakkuk.net/ and click on the "Board Games and More" file. It is recommended that you use white cardstock to print the available title, heading cards, and spinners. (Gameboards with start to finish paths are also available for printing.)

Cut out the pieces and tape the title towards the top of a sheet of poster board. Tape the heading cards to lunch bags or to inexpensive colored gift bags as shown in the illustration. Then tape the lunch bags to the poster board. To prepare the spinners, you will need two brads and two large paperclips. Reproduce the spinners on cardstock and cut out. Put a brad through the end of a paperclip and push it through the middle of the spinner. Turn the spinner over to fasten the brad to the back.

Most of the social studies or science lessons in the books include unit questions to use with this game. Reproduce the questions on cardstock and cut them apart.

INSTRUCTIONS ON HOW TO PLAY: Decide if you want students to play in teams or individually. Then let a child spin the spinner. If the spinner lands on *Seasons and Other Science Themes*, he or she draws a unit question card from the corresponding bag. If the child answers the question correctly, let him hold on to the card until the end of the game. The child or team with the most cards is the winner.

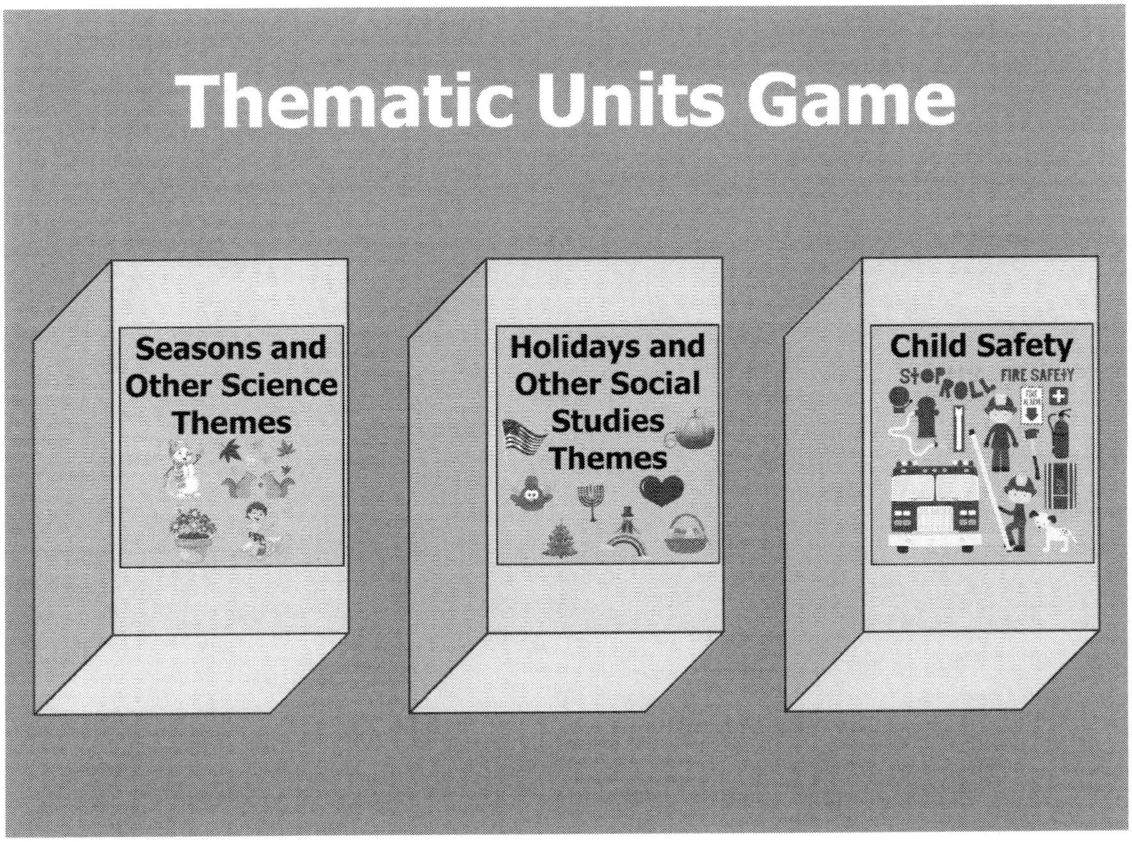

Name: ______________________________

Date: ______________________________

Writing Paper

Scientist: ______________________________

Experiment: ______________________________

Evaluation of Science Experiments

Participation

- Completed the experiment (10 points): _____
- Completed more than ½ of the experiment (7½ points): _____
- Completed ½ of the experiment (5 points): _____
- Completed less than ½ of the experiment (2½ points): _____
- Never began the experiment (0 points): _____

Comprehension

- Answered 100% of the corresponding worksheet questions appropriately (10 points): _____
- Answered more than ½ of the corresponding worksheet questions appropriately (7½ points): _____
- Answered ½ of the corresponding worksheet questions appropriately (5 points): _____
- Answered less than ½ of the corresponding worksheet questions appropriately (2½ points): _____
- Answered 100% of the corresponding worksheet questions inappropriately (0 points): _____

Total Points: _____ (Scores are added and multiplied by 5.)

Name: ______________________________

Spelling Test / Pretest

1. ____________________
2. ____________________
3. ____________________
4. ____________________
5. ____________________
6. ____________________
7. ____________________
8. ____________________
9. ____________________
10. ____________________
11. ____________________
12. ____________________

Bonus words: ______________________________

____

____

____

Lesson 1

Theme: The Summer Season
Sound: short *u* /ŭ/
Spelling Patterns: *u*, *o*

Spelling Words	Alphabetized
1. July	______________
2. sun	______________
3. sunny	______________
4. fun	______________
5. summer	______________
6. jump	______________
7. run	______________
8. lunch	______________
9. bugs	______________
10. August	______________
11. month	______________
12. shovel	______________

Inflectional endings: jumps, jumping, jumped

Bonus compound words: summertime, sunburn, sunglasses, sunscreen, sunshine, suntan

Note: The initial letter of *July* and *August* must be capitalized.

Thematic Activities for Lesson 1
(The Summer Season)

Thematic objectives: Students will understand that summer is one of the four seasons and will be able to identify weather conditions and activities that occur in summer.

Key Questions for Discussion

- Summer is one of the four seasons. What are the four seasons? (*winter, spring, summer, fall*)
- What happens during the summer season? (*It is very hot and sunny. Most schools are closed. People wear shorts and go swimming during the summertime.*)
- What are the summer months? (*Part of June, all of July and August, part of September*)

Activities, Supplies, and Preparations for this Lesson

- **Spelling Patterns of Sounds Heard in the English Language:** Display words with this week's sound on the wall display. (See pages 26-27 for details.)
- **Reproducible materials:** Prepare the reproducible materials provided with this lesson, including the question cards and the spelling word cards
  (Supplies: Computer paper, cardstock, magnetic tape)
- **Books and videos:** Read books and/or watch YouTube videos about seasons and summer. Have Fun Teaching has a cute video available on YouTube titled "Seasons Song" that can help students learn about the four seasons. (For any books you read, identify the title, the author, and the illustrator. Also, examine the front and back cover of the book before making any predictions about the story or discussing what it is about.)
- **Crafts:** Choose a craft activity for making suns from online. Decorate the room with the suns students make.
- **Writing/Handwriting assignment**
  Younger Students: Make a copy of the reproducible writing paper provided on page 30 and position it under a Smart Document Camera or a similar device. Discuss the correct stroke for each letter as you copy the incomplete sentence below (or one of your own choosing) onto your paper. Students should follow along with you, copying the letters on papers of their own. (This will help them to remain on task and to finish the handwriting assignment in unison.) They can then finish the sentence, add additional sentences, and include appropriate illustrations.

  My favorite thing about the summer season is...

  Older Students: Older students can write about the topic below. Let them copy the question and respond to it in their writing journals. They may also want to add appropriate illustrations.

  What is your favorite thing about the summer season and why?

Context Clues

Directions: Fill in each blank with a spelling word that makes sense in the context of the sentence. (Cross out the words from the list as you use them.) Circle the letter in the key words that makes the short *u* sound (*u*, *o*).

Spelling words: July, sun, sunny, fun, summer, jump, run, lunch, bugs, August, month, shovel

1. ____________ is a time when it is hot outside.
2. Most days are ____________.
3. The first ____________ of summer is June.
4. After June is ____________.
5. ____________ is the last full month of summer.
6. We mostly just have ____________ in the summer because school is out.
7. We might ____________ into a swimming pool.
8. We might use the ____________ Mom got us to dig in the sand.
9. We look for ____________ crawling around.
10. We eat ____________ outside.
11. We ____________ through the sprinkler.
12. We have all sorts of fun while the ____________ is shining.

Spelling Patterns of Sounds Heard in Our Spelling Words

Directions: Fill in the missing letters that make the short *u* sound /ŭ/.

- Use a **blue** colored pencil for ***u***.
- Use a **red** colored pencil for ***o***.

J _ l y	s _ m m e r	b _ g s
s _ n	j _ m p	A u g _ s t
s _ n n y	r _ n	m _ n t h
f _ n	l _ n c h	s h _ v e l

Directions: Record spelling words whose short *u* sound is spelled *u* under the **"u"** headings and words whose /ŭ/ sound is spelled *o* under the **"o"** heading.

u	u	o
______	______	______
______	______	______
______	______	______
______	______	______
______	______	______

Spelling Words (Short *u* /ŭ/ sound)

A	U	G	U	S	T	A	N	E	R	U	N
U	K	L	I	L	K	S	U	T	F	B	X
S	U	M	D	M	T	S	H	O	V	E	L
J	T	O	M	S	T	Y	V	D	X	H	Z
U	P	N	C	U	N	R	D	J	Z	D	S
L	I	T	F	N	D	O	A	L	L	N	U
Y	A	H	Q	V	K	M	W	C	U	S	N
A	F	N	B	U	G	S	P	K	N	T	N
G	S	L	W	E	G	H	B	C	C	W	Y
X	F	Y	Y	J	U	M	P	H	H	Q	C
I	U	Q	E	I	P	X	Y	V	F	M	R
Q	N	L	F	B	Z	S	U	M	M	E	R

The Summer Season

1. July
2. sun
3. sunny
4. fun
5. summer
6. jump
7. run
8. lunch
9. bugs
10. August
11. month
12. shovel

Homophones

son

My **son** likes the carp streamers.

A son is the boy child of a mother and father.

sun

The **sun** feels hot today.

The sun is the bright light in the sky that shines during the day.

My **son** and I are having fun in the **sun**.

Underline the word that refers to the boy child of a mother and father.
Circle the word that refers to the bright light in the sky that shines during the day.

Compound Words

Directions: When two or more root words are used together to form one word, it is called a **compound word** (summertime). Complete each sentence by writing a word to form a compound in the blank space. The word must make sense in the context of the sentence.

1. Michael is wearing sun__________ to protect his eyes from the sun__________.
   (burn, glasses, screen, shine, tan)

2. Mandi put on some sun__________ to keep from getting a sun__________.
   (burn, glasses, screen, shine, tan)

3. Shana is laying out in the sun because she wants a sun__________.
   (burn, glasses, screen, shine, tan)

Weekly Oral Language

1. Correct the asking sentence.

   can we eat lunch outside today

2. Correct the telling sentence.

   we can eat lunch outside today

3. What is the subject of the telling sentence?

   ____

4. Read the *telling sentence* again and identify the helping verb and the verb (action word).

   Helping Verb: ______________________

   Verb: ______________________________

5. Write your own sentence about this week's theme.

   ____

Instruction: *Teaching Children about Capitalization, Punctuation, Apostrophe s, and Contractions (Language Book 2)* by Habakkuk Educational Materials can be purchased separately to give students the knowledge they need to complete this worksheet. After students have completed the worksheet and turned it in, complete a copy of it on your classroom screen.

Parts of Speech

Directions: Identify what parts of speech are used in the sentence below and write them in the spaces provided. Then identify the subject (naming part) and predicate (action part) of the sentence and write them in the spaces provided.

Parts of speech: Noun marker, noun, pronoun, verb, helping verb, adjective, adverb, preposition, conjunction, interjection

We	have	fun	in	summer.
__________	__________	__________	__________	__________

Subject (Naming Part)

____

Predicate (Action Part)

____

Instruction: *Teaching Children about Parts of Speech (Language Book 3)* by Habakkuk Educational Materials can be purchased separately to help students quickly learn the parts of speech. After students have completed this worksheet and turned it in, complete a copy of it on your classroom screen.

Schwa (Short *U* Sound /Ŭ/)

Directions: Write letters in the spaces provided to form words with the key sounds. Call on students to sound out the words Then use one or more of the words in a sentence.

C<u>o</u>me　　　　**D<u>o</u>n**e

______o______　　______o______

______o______　　______o______

L<u>o</u>ve

______o______

______o______

____

____

Rule: The vowel *o* often makes the short *u* sound (a schwa sound), /ə/, when it's followed by *m*, *n*, or *v* (and is referred to as a "scribal *o*").

Examples: c<u>o</u>**m**e, fr<u>o</u>**m**, d<u>o</u>**n**e, t<u>o</u>**n**, ab<u>o</u>**v**e, l<u>o</u>**v**e

Name: ______________________

Addition and Subtraction Facts

Directions

1. Count the ants above the first space and write the addend.
2. Count the ants above the second space and write the addend.
3. Count all the ants and write the sum after the equal sign.

______ + ______ = ______

Directions

1. Count the ants above the first space and write the number.
2. Cross out 3 ants.
3. Count the number of ants left and write the answer after the equal sign.

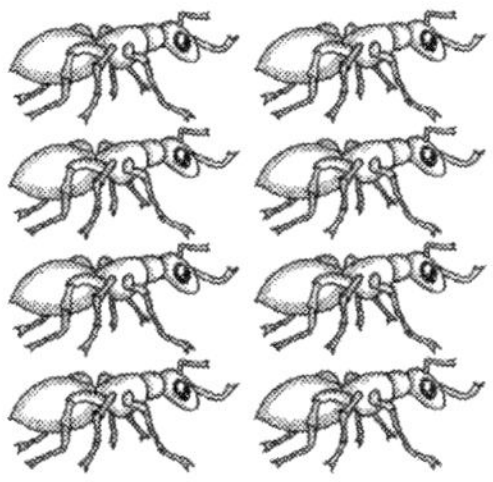

______ − 3 = ______

Addition and Subtraction Word Problems

Directions: Color and cut out the pictures at the bottom of the page. Read each sentence and use the cutouts to help solve the problem. Write the addition or subtraction fact in the spaces provided.

1. There were 7 scoops of ice cream on the cone. Two fell off. How many scoops of ice cream are left on the cone?

   ___________ = _____

2. There were 2 scoops of ice cream on the cone. The worker in the ice cream shop put on four more scoops. How many scoops are on the cone?

   ___________ = _____

3. There were 9 scoops of ice cream on the cone. Five fell off. How many scoops of ice cream are left on the cone?

   ___________ = _____

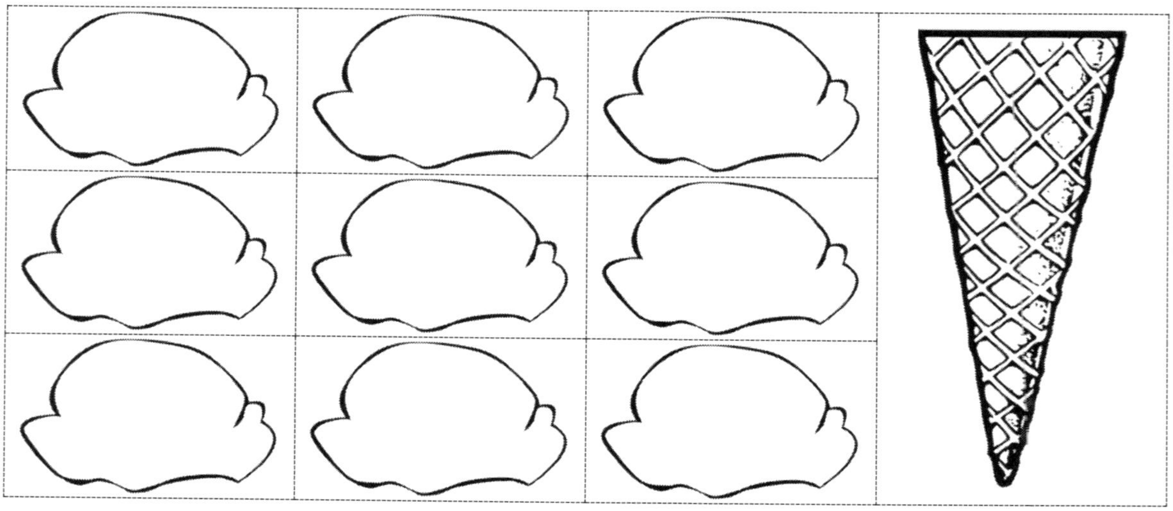

QUESTIONS FOR THE THEMATIC UNITS GAME

Please refer to pages 28-29 of this book for instructions on how to use the unit questions.

Summer is one of the four seasons. What are the four seasons? *Lesson 1 (Science Themes)*	What happens during the summer season? *Lesson 1 (Science Themes)*
What are the summer months? *Lesson 1 (Science Themes)*	

LESSON 1 TEST (The Summer Season)

1. What are the four seasons?
   - ◯ Sunday, Monday, Tuesday, and Wednesday
   - ◯ September, October, November, and December
   - ◯ Winter, spring, summer, and fall

2. What is the weather usually like in summer?
   - ◯ Cold and snowy
   - ◯ Hot and sunny
   - ◯ Freezing and cloudy

3. What is a summertime activity?
   - ◯ Swimming outside
   - ◯ Eating Thanksgiving dinner
   - ◯ Building a snowman

4. What are the summer months?
   - ◯ December, January, and February
   - ◯ March, April, and May
   - ◯ June, July, and August

Note: These cards can be used for the spelling word games listed on pages 21-22.

J<u>u</u>ly	s<u>u</u>n
s<u>u</u>nny	f<u>u</u>n
s<u>u</u>mmer	j<u>u</u>mp

Note: These cards can be used for the spelling word games listed on pages 21-22.

r<u>u</u>n	l<u>u</u>nch
b<u>u</u>gs	Aug<u>u</u>st
m<u>o</u>nth	sh<u>o</u>vel

Lesson 2

Theme: Visiting the Ocean
Sound: /sh/
Spelling Pattern: *sh*

Spelling Words	Alphabetized
1. ship	______________
2. shark	______________
3. shell	______________
4. shine	______________
5. shallow	______________
6. shore	______________
7. shade	______________
8. shorts	______________
9. shovel	______________
10. shadow	______________
11. fish	______________
12. splash	______________

Inflectional endings: splashes, splashing, splashed; shallower, shallowest

Bonus compound words: clamshell, seashell, seashore, starfish, sunshade, sunshine

Extra challenge words: ocean, lotion

Thematic Activities for Lesson 2 (Visiting the Ocean)

Thematic objectives: Students will learn about the beach and be able to identify the five oceans (*the Pacific, Atlantic, Indian, Arctic, and Southern*). They will learn that the Pacific Ocean is the largest ocean and that the earth has more water than land.

Key Questions for Discussion

- What are the five oceans? (*the Pacific, Atlantic, Indian, Arctic, and Southern*)
- What is another name for the Southern Ocean, which is near to the continent Antarctica? (*Antarctic Ocean*)
- How much of the earth is covered with water? (*more than half or 71%—the earth has more water than land*)
- What is the largest ocean? (*Pacific*)

Activities, Supplies, and Preparations for this Lesson

- **Spelling Patterns of Sounds Heard in the English Language:** Display words with this week's sound on the wall display. (See pages 26-27 for details.)
- **Reproducible materials:** Prepare the reproducible materials provided with this lesson, including the question cards and the spelling word cards (Supplies: Computer paper, cardstock)
- **Books and videos:** Read books and/or watch YouTube videos about oceans and the beach. (For any books you read, identify the title, the author, and the illustrator. Also, examine the front and back cover of the book before making any predictions about the story or discussing what it is about.)
- **Turning Your Classroom into an Ocean:** Pinterest has plenty of ideas online for turning your classroom into an underwater adventure. Blue, plastic tablecloths from the Dollar Tree can be hung on the walls and used to dim the ceiling lights. Jellyfish can be made and hung from the ceilings, fish crafts completed by students can be attached to walls, and inflatable fish and other sea creatures can be provided.
- **Fish crafts:** Choose a fish craft activity for kids from online and display their creations in your classroom.
- **Math Center Activity:** Follow the instructions on page 58.
  (Supplies: Cardstock, miniature seashells, Elmer's glue)

- **Subtraction Word Problems for Centers:** Follow the instructions on page 60. Print a free clipart image of a great white shark from the internet and tape it to a blue paper sack or to a shoebox wrapped in blue wrapping paper to represent the ocean. If you use a shoebox, an X-Acto knife can be used to cut a slit in the box where the shark's mouth is.
(Supplies: Clipart picture of a great white shark; a blue paper sack or a shoebox wrapped in blue wrapping paper; X-Acto knife)
- **The Five Oceans:** Display a simple continents and oceans colored map poster on your classroom wall and let students identify the seven continents (*North America, South America, Africa, Australia, Asia, Antarctica, and Europe*) and the five oceans (*Pacific, Atlantic, Indian, Arctic, and Southern*) daily throughout the remainder of the year. (Note that the Southern Ocean is also known as the Antarctic Ocean.)
(Supplies: A simple continents and oceans colored map poster that has been laminated)
- **Thematic Units Game for Oceans:** Reproduce the unit questions and the spinner from page 61 on cardstock and cut apart the pieces. Use wall-safe tape to attach the questions to different oceans on a laminated poster map. As a variation or in addition to using the unit questions, the spelling word cards can also be used with the game. To prepare the spinner, put a brad through the end of a large paperclip and push it through the middle of the spinner. Turn the spinner over to fasten the brad to the back. A student spins the paperclip and chooses a question to answer or a spelling word to read from the ocean it lands on.
(Supplies: The laminated map from the previous activity, cardstock, a brad, a large paperclip, wall-safe tape)
- **Writing/Handwriting assignment**
Younger Students: Make a copy of the reproducible writing paper provided on page 30 and position it under a Smart Document Camera or a similar device. Discuss the correct stroke for each letter as you copy the incomplete sentence below (or one of your own choosing) onto your paper. Students should follow along with you, copying the letters on papers of their own. (This will help them to remain on task and to finish the handwriting assignment in unison.) They can then finish the sentence, add additional sentences, and include appropriate illustrations.

  If I was going to the beach today, the first thing I would do is...

  Older Students: Older students can write about the topic below. Let them copy the question and respond to it in their writing journals. They may also want to add appropriate illustrations.

  If you were going to the beach today, what would you do there?

Context Clues

Directions: Fill in each blank with a spelling word that makes sense in the context of the sentence. (Cross out the words from the list as you use them.) Circle the letters in the key words that make the /sh/ sound.

Spelling words: ship, shark, shell, shine, shallow, shore, shade, shorts, shovel, shadow, fish, splash

1. I'm wearing ____________ because I'm going to the beach.
2. Mom is bringing the beach umbrella to give us ____________ from the sun.
3. The sun likes to ____________ down on the beach!
4. There's a lot to do on the sandy ____________.
5. The sun helps me to see everybody's ____________.
6. I can use my ____________ to dig in the sand.
7. I can also look for a crab ____________.
8. From the shore, I can see when a ____________ sails by.
9. I can also see people on boats looking for ____________.
10. One time I saw a fin that may have belonged to a ____________.
11. When I get too hot, I get in the ____________ part of the water.
12. I like it when the waves ____________ me.

Spelling Words (/sh/ sound)

K	C	S	V	B	S	H	A	L	L	O	W
U	G	H	I	S	K	S	U	T	F	B	X
S	U	O	D	P	T	F	R	O	E	S	V
H	T	R	M	L	T	I	V	S	X	H	Z
E	P	T	C	A	N	S	D	H	Z	A	A
L	I	S	F	S	D	H	A	A	M	D	L
L	A	D	Q	H	K	M	W	D	Z	O	S
A	F	N	Z	Q	F	H	P	E	G	W	H
S	H	O	R	E	G	H	B	C	R	W	A
X	G	Y	Y	W	S	H	O	V	E	L	R
I	O	Y	R	O	P	X	Y	V	F	M	K
S	H	I	P	B	S	H	I	N	E	J	X

Visiting the Ocean

1. ship
2. shark
3. shell
4. shine
5. shallow
6. shore
7. shade
8. shorts
9. shovel
10. shadow
11. fish
12. splash

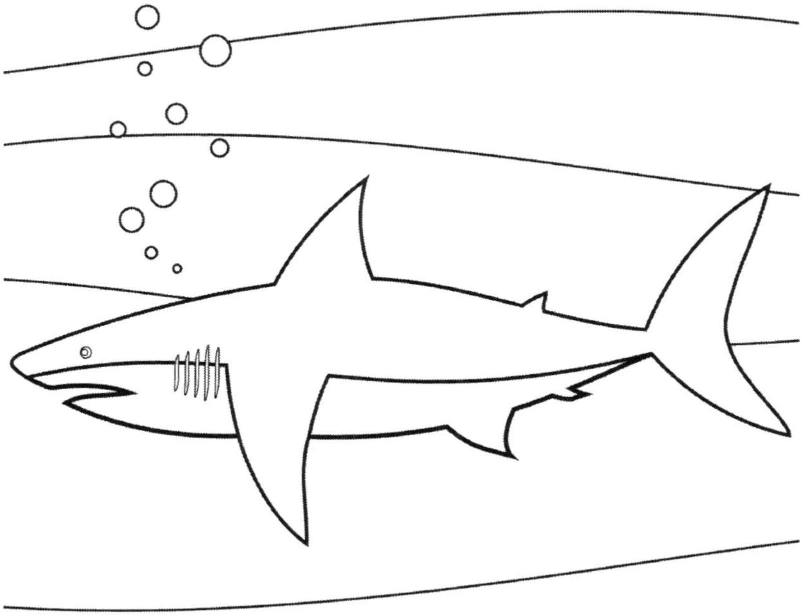

Compound Words

Directions: When two or more root words are used together to form one word, it is called a **compound word** (clamshell). Form five compound words by adding a spelling word to the words *sea*, *star*, and *sun*.

Spelling words: fish, shade, shells, shine, shore

sea________ sea________ star________

sun________ sun________

Directions: Fill in each blank with one of the compound words that makes sense in the context of the sentence.

Today we went to the _______________ to look for _______________ on the sand. The _______________ was making us hot, and we forgot to bring our _______________, so we got into the shallow part of the ocean to cool off. We spotted a fish with five pointed legs. Mom said it was a _______________.

Weekly Oral Language

1. Correct the asking sentence.

   What are the five oceans

2. Correct the telling sentence.

   the five oceans are the pacific, atlantic, indian, arctic, and southern

3. What is the subject of the telling sentence?

   ____

4. Read the *telling sentence* again and identify the verb.

   ____

5. Write your own sentence about this week's theme.

   ____

Instruction: *Teaching Children about Capitalization, Punctuation, Apostrophe s, and Contractions (Language Book 2)* by Habakkuk Educational Materials can be purchased separately to give students the knowledge they need to complete this worksheet. After students have completed the worksheet and turned it in, complete a copy of it on your classroom screen.

Parts of Speech

Directions: Identify what parts of speech are used in the sentence below and write them in the spaces provided. Then identify the subject (naming part) and predicate (action part) of the sentence and write them in the spaces provided.

Parts of speech: Noun marker, noun, pronoun, verb, helping verb, adjective, adverb, preposition, conjunction, interjection

We	**saw**	**a**	**huge**	**shark!**
________	________	________	________	________

Subject (Naming Part)

____

Predicate (Action Part)

____

Instruction: *Teaching Children about Parts of Speech (Language Book 3)* by Habakkuk Educational Materials can be purchased separately to help students quickly learn the parts of speech. After students have completed this worksheet and turned it in, complete a copy of it on your classroom screen.

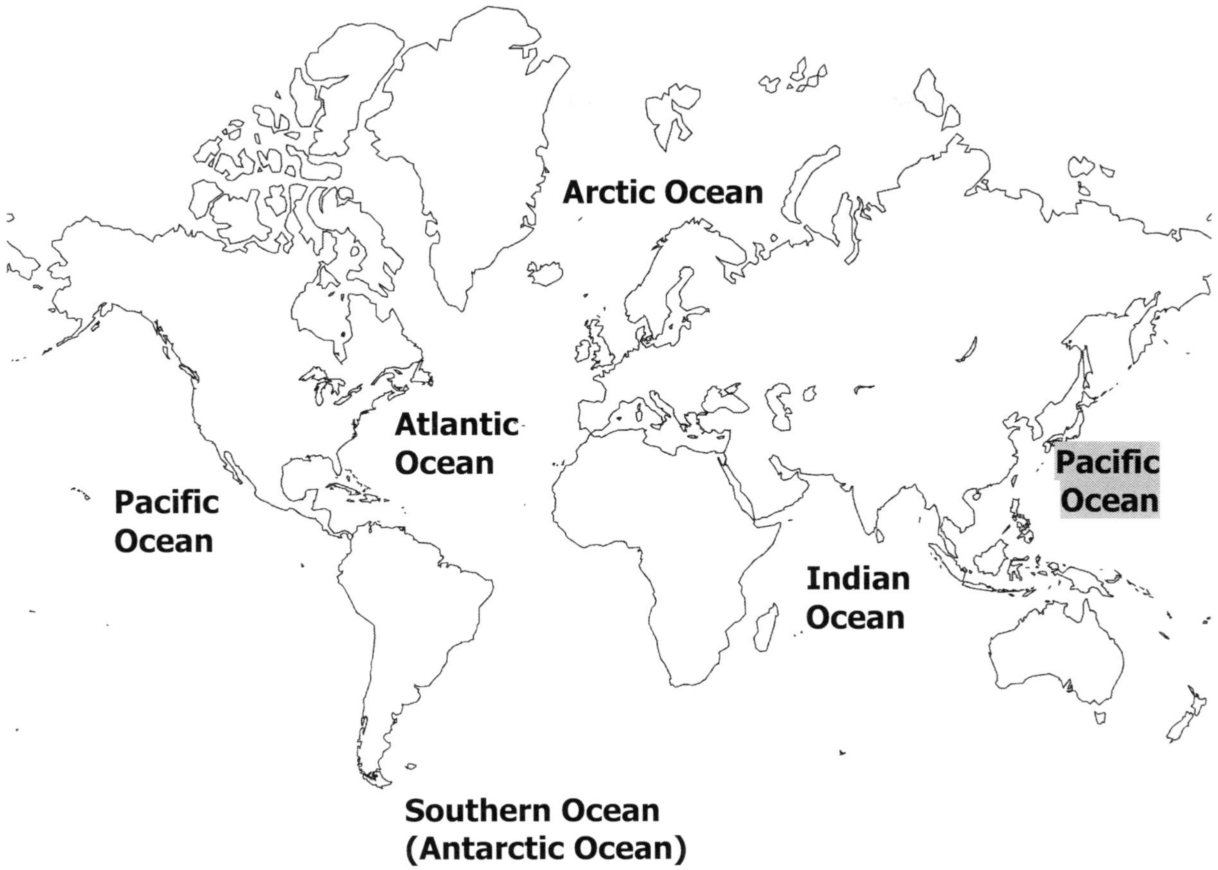

Directions: The map shows the seven continents and five oceans on our planet. The oceans are labeled in the top map. Notice that there is more water on Earth than land. Ocean water does cover over half of the earth (71% of Earth is covered with water). Black dots were used to mark the five oceans on the map below. Label the dots with one of the oceans (*Pacific*, *Atlantic*, *Indian*, *Arctic*, or *Southern*). The reason the Pacific Ocean is recorded on two areas of the map above is because the world is round. The Pacific is the largest ocean. Note also that the Southern Ocean, which is near Antarctica, is also called the Antarctic Ocean.

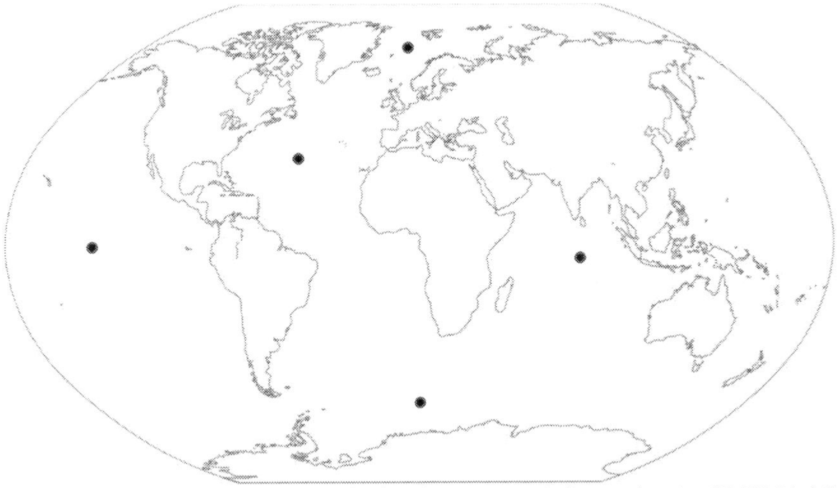

Math Center Directions: Reproduce this page on cardstock. Students record a 1-digit number (or the name of a number, such as *one*) in the spaces before the equal sign, glue the corresponding number of miniature seashells above the addends, and then count the seashells to find the sum.

addend		addend		sum
___	+	___	=	___

addend		addend		sum
___	+	___	=	___

addend		addend		sum
___	+	___	=	___

Name: ______________________________

Addition and Subtraction Facts

Directions

1. Count the fish above the first space and write the addend.
2. Count the fish above the second space and write the addend.
3. Count all the fish and write the sum after the equal sign.

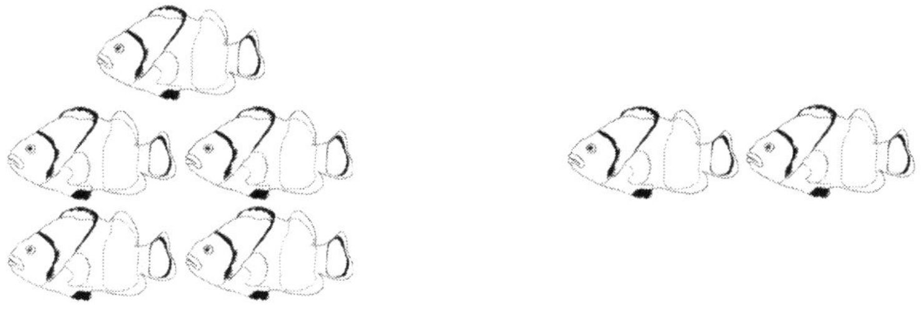

________ + ________ = ________

Directions

1. Count the shells above the first space and write the number.
2. Cross out 5 shells.
3. Count the number of shells left and write the answer after the equal sign.

________ − 5 = ________

Subtraction Word Problems for Centers

Teacher Instructions: Print a free clipart image of a great white shark from the internet and tape it to a blue paper sack or to a shoebox wrapped in blue wrapping paper to represent the ocean. (If you use a shoebox, an X-Acto knife can be used to cut a slit in the box where the shark's mouth is. Let students color and cut out the pictures at the bottom of this page. For the first problem, they would count the number of fish there were originally, place the number of fish eaten by the shark in the sack or box, and then count how many are left. (You might also want to make up your own problems for a group-time activity.)

1. A great white shark came upon a group of **5** fish. If he ate **3** of them, how many fish got away?

   _____ – _____ = _____

2. A great white shark came upon a group of **4** fish. If he ate **1** of them, how many fish got away?

   _____ – _____ = _____

3. There were **5** shells on the shore. The ocean washed **4** away. How many shells are left on the shore?

   _____ – _____ = _____

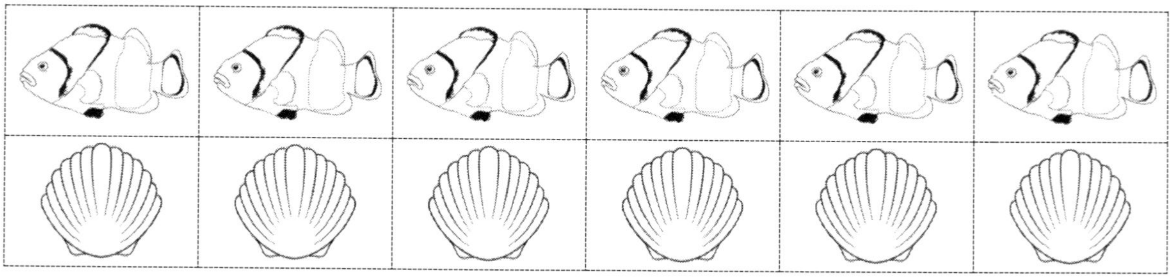

QUESTIONS FOR THE THEMATIC UNITS GAME

Please refer to pages 28-29 of this book for instructions on how to use the unit questions.

What are the five oceans? *Lesson 2 (Science Themes)*	What is another name for the Southern Ocean, which is near to the continent Antarctica? *Lesson 2 (Science Themes)*
How much of the earth is covered with water? *Lesson 2 (Science Themes)*	What is the largest ocean? *Lesson 2 (Science Themes)*

Directions: The spinner and unit questions can be used for the thematic units game described on page 51.

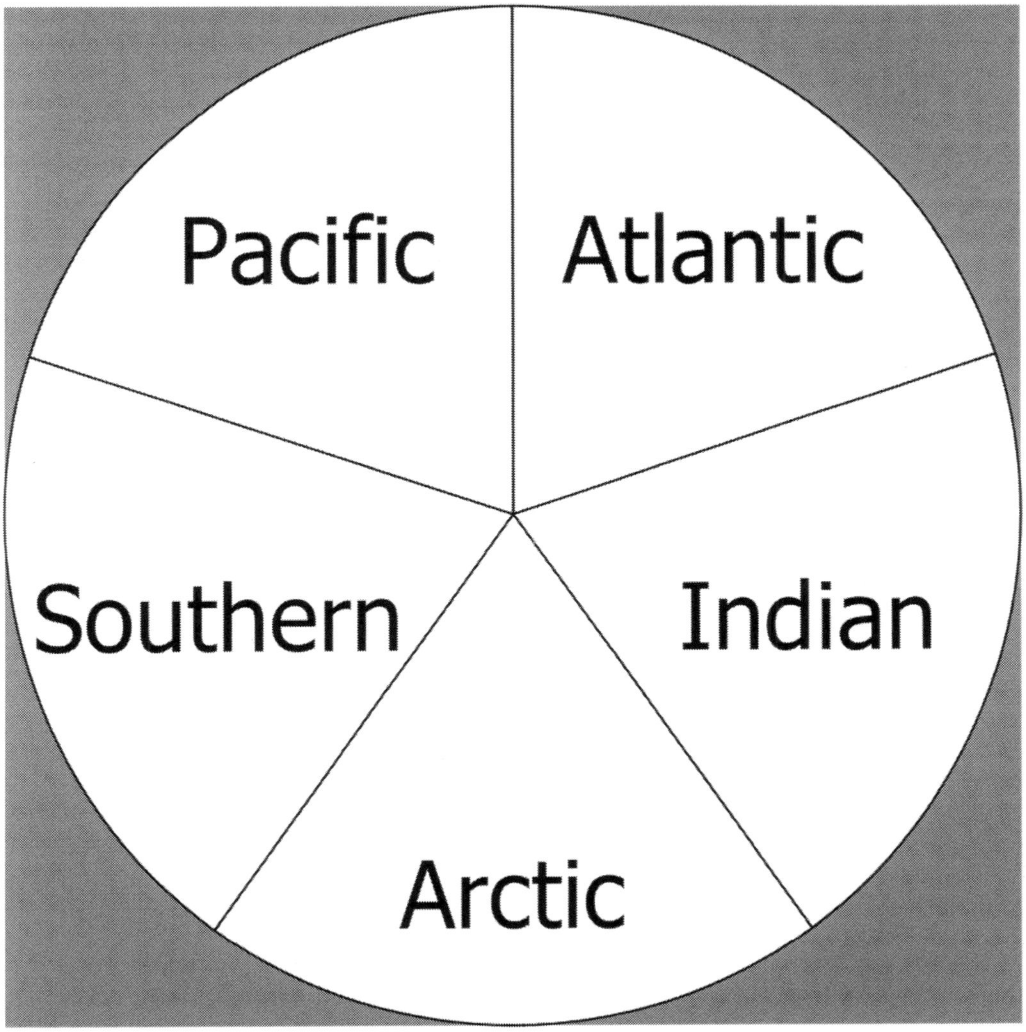

LESSON 2 TEST (Visiting the Ocean)

1. What are the five oceans?

   ○ Pacific

   ○ Delaware

   ○ Atlantic

   ○ Indian

   ○ Arctic

   ○ Rio Grande

   ○ Southern

2. What is another name for the Southern Ocean, which is near to the continent Antarctica?

   ○ Antarctic

   ○ Nile

   ○ Rio Grande

3. Which of the following statements is true?

   ○ The earth has more land than water.

   ○ The earth has more water than land.

   ○ The earth has an equal amount of land and water.

4. How much of the earth is covered with water?

   ○ About 25%

   ○ About 50%

   ○ About 71%

5. What is the largest ocean?

- ○ Pacific
- ○ Atlantic
- ○ Indian

6. Identify the five oceans (Arctic, Atlantic, Indian, Pacific, Southern).

This illustration from World map longlat by Thesevenseas is licensed under CC BY-SA 4.0.
http://www.supercoloring.com/coloring-pages/world-map?tag=63216

Note: These cards can be used for the spelling word games listed on pages 21-22.

<u>sh</u>ip	<u>sh</u>ark
<u>sh</u>ell	<u>sh</u>ine
<u>sh</u>allow	<u>sh</u>ore

Note: These cards can be used for the spelling word games listed on pages 21-22.

shade	shorts
shovel	shadow
fish	splash

Lesson 3

Theme: Camping Trip
Sound: /u/ sound in *look*
Spelling Patterns: *oo, o, u*

Spelling Words	Alphabetized
1. took	________________
2. brook	________________
3. cook	________________
4. hook	________________
5. look	________________
6. good	________________
7. wood	________________
8. woods	________________
9. wolf	________________
10. wolves	________________
11. pull	________________
12. put	________________

Inflectional endings: looks, looking, looked

Bonus compound words: firewood, fishhook

Context Clues

Directions: Fill in each blank with a spelling word that makes sense in the context of the sentence. (Cross out the words from the list as you use them.) Circle the letters in the key words that make the /u/ sound heard in *look* (*oo*, *o*, *u*).

Spelling words: took, brook, cook, hook, look, good, wood, woods, wolf, wolves, pull, put

1. Dad ____________ me camping last weekend.
2. We set up our tent in the ____________.
3. Then we found a ____________ to fish in.
4. Dad showed me how to bait my fishing ____________!
5. Then I ____________ the fishing line in the water and waited.
6. When a fish took the bait, Dad yelled, "____________ it in!"
7. After fishing, Dad and I gathered some ____________ to make a campfire with.
8. He showed me how to ____________ the fish.
9. It tasted really ____________!
10. That night, we heard some ____________ howling.
11. They ____________ a lot like dogs.
12. Dad laughed when I told him I wanted a pet ____________.

Spelling Word Riddles

Discussion: All spelling words this week have the /u/ sound heard in *look* and can be spelled as ***oo*** as in *hook*, ***o*** as in *wolf*, or ***u*** as in *put*. (Help students hear and pronounce the /u/ sound by first saying the sound and then a basic word, such as *look*.)

Directions: Reproduce the next two pages on light blue or yellow cardstock as instructed. Cut out the pieces and attach a magnetic strip to the back of each piece. Line up the yellow pieces horizontally towards the top of a magnetic board. Position the light blue strips vertically below them. (See the illustration on page 15.) Read a riddle and call on a student to identify the correct spelling word. For the first sentence, a student would identify the word *cook* and then move the yellow magnetic strip with the letters ***oo*** between the light blue letters ***c*** and ***k*** to form the word ***cook***.

1. In this word, the /u/ sound is spelled *o-o*. If you catch a fish, you can't eat it until you do this. (*Cook*)
2. In this word, the /u/ sound is spelled *o-o*. This is the opposite of *bad*. (*Good*)
3. In this word, the /u/ sound is spelled *o*. You might see this animal in the woods. (*Wolf*)
4. In this word, the /u/ sound is spelled *o-o*. If you go fishing, this is what you will put your bait on. (*Hook*)
5. In this word, the /u/ sound is spelled *o-o*. You need this to start a campfire. (*Wood*)
6. In this word, the /u/ sound is spelled *o*. This is the plural of *wolf*. (*Wolves*)
7. In this word, the /u/ sound is spelled *o-o*. This is a small stream. (*Brook*)
8. In this word, the /u/ sound is spelled *o-o*. This is what you do with binoculars. (*Look*)
9. In this word, the /u/ sound is spelled *u*. It begins with *p* and rhymes with *foot*. (*Put*)
10. In this word, the /u/ sound is spelled *o-o*. It is the past tense of *take*. (*Took*)
11. In this word, the /u/ sound is spelled *o-o*. This is a place with a lot of trees that people go camping in. (*Woods*)
12. In this word, the /u/ sound is spelled *u*. If you catch a big fish, you will probably have to do this to bring it in. (*Pull*)

Directions: Use light blue cardstock to reproduce this page. Cut out the pieces and attach magnetic strips to the back of each piece. (See the previous page for further instructions.)

t	k	br	k
c	k	h	k
l	k	g	d
w	d	w	ds
w	lf	w	lves
p	ll	p	t

Directions: Use yellow cardstock to reproduce this page. Cut out the pieces and attach magnetic strips to the backs.

oo	oo	oo	oo	oo			
oo	oo	oo	oo	oo			
o	o	o	o	u	u	u	u

Spelling Patterns of Sounds Heard in Our Spelling Words

Directions: Fill in the missing letters that make the /u/ sound heard in <u>*look*</u>.

- Use a **blue** colored pencil for ***oo***.
- Use a **red** colored pencil for ***o***.
- Use a **green** colored pencil for ***u***.

t _ _ k	l _ _ k	w _ l f
b r _ _ k	g _ _ d	w _ l v e s
c _ _ k	w _ _ d	p _ l l
h _ _ k	w _ _ d s	p _ t

Directions: Record spelling words whose /u/ sound is spelled *oo* under the **"oo"** headings, words whose /u/ sound is spelled *o* under the **"o"** heading, and so forth.

oo	**oo**	**o**	**u**

Spelling Words (/u/ sound in *look*)

P	U	T	V	B	T	A	N	P	U	L	L
U	K	L	I	L	O	U	S	K	F	B	X
S	U	C	D	M	O	O	W	O	O	D	S
C	T	O	M	H	K	Y	V	D	X	H	Z
K	P	O	C	H	N	R	D	J	B	D	A
B	I	K	F	J	D	O	A	L	R	N	L
W	D	A	Q	G	O	O	D	C	O	S	J
O	F	N	Z	Q	F	H	P	K	O	T	E
O	S	L	O	O	K	H	B	C	K	W	H
D	G	Y	Y	W	M	E	P	H	I	Q	O
I	R	Q	E	I	P	W	O	L	F	M	O
W	O	L	V	E	S	U	I	J	P	J	K

Camping Trip

1. took
2. brook
3. cook
4. hook
5. look
6. good
7. wood
8. woods
9. wolf
10. wolves
11. pull
12. put

Homophones

wood

We gathered **wood**
for the campfire.

Wood comes from a tree.

would

Would you like to color
the question mark?

Use *would if you are NOT talking about something that comes from a tree.*
Would is sometimes a question word.

Would you please gather
more **wood** for the fire?

Underline the word that comes from a tree.
Circle the question word.

Weekly Oral Language

1. Correct the asking sentence.

   what are you eating

2. Correct the telling sentence.

   we are eating fish

3. What is the subject of the telling sentence?

   ____

4. Read the *telling sentence* again and identify the helping verb and the verb (action word).

   Helping Verb: ______________________

   Verb: ______________________________

5. Write your own sentence about this week's theme.

   ____

Instruction: *Teaching Children about Capitalization, Punctuation, Apostrophe s, and Contractions (Language Book 2)* by Habakkuk Educational Materials can be purchased separately to give students the knowledge they need to complete this worksheet. After students have completed the worksheet and turned it in, complete a copy of it on your classroom screen.

Parts of Speech

Directions: Identify what parts of speech are used in the sentence below and write them in the spaces provided. Then identify the subject (naming part) and predicate (action part) of the sentence and write them in the spaces provided.

Parts of speech: Noun marker, noun, pronoun, verb, helping verb, adjective, adverb, preposition, conjunction, interjection

I	caught	a	big	fish!
__________	__________	__________	__________	__________

Subject (Naming Part)

____

Predicate (Action Part)

____

Instruction: *Teaching Children about Parts of Speech (Language Book 3)* by Habakkuk Educational Materials can be purchased separately to help students quickly learn the parts of speech. After students have completed this worksheet and turned it in, complete a copy of it on your classroom screen.

Name: ______________________________

Addition and Subtraction Facts

Directions

1. Count the fish above the first space and write the addend.
2. Count the fish above the second space and write the addend.
3. Count all the fish and write the sum after the equal sign.

______ + ______ = ______

Directions

1. Count the wolves above the first space and write the number.
2. Cross out 3 wolves.
3. Count the number of wolves left and write the answer after the equal sign.

______ − 3 = ______

FISHING GAME: Cut out the cards and attach a paperclip to each word. Prepare a magnetic fishing pole by tying one end of a string of yarn to a stick or dowel rod and the other to a magnet with a hole in it. Let each child "fish" for a word and instruct him/her to name the letters of a caught word before reading it.

took	brook
cook	hook
look	good

Note: These cards can be used for the spelling word games listed on pages 21-22.

wood	woods
wolf	wolves
pull	put

Lesson 4

Theme: Space
Sound: /ü/ sound in *moon*
Spelling Patterns: *oo, ew, u, ui, ough*

Note to teacher: Try to have a solar system poster displayed throughout the week.

Spelling Words	Alphabetized
1. moon	______________
2. shooting	______________
3. zoom	______________
4. crew	______________
5. flew	______________
6. view	______________
7. lunar	______________
8. Jupiter	______________
9. Neptune	______________
10. Pluto	______________
11. spacesuit	______________
12. through	______________

Inflectional endings: zooms, zooming, zoomed

Bonus compound words: throughout, full moon, half-moon, shooting star

Note: The initial letter of *Jupiter*, *Neptune*, and *Pluto* must be capitalized.

Thematic Activities for Lesson 4
(Space)

Thematic objectives: Students will be able to identify the eight planets in our solar system, their distance (closest to furthest) from the sun, and the smallest and largest planet. They will also learn about the Apollo 11 spaceflight, about shooting stars, the Milky Way Galaxy, and about the International Space Station (ISS).

Key Questions for Discussion

- Who was the crew of the Apollo 11 spaceflight in 1969? (*Neil Armstrong, Buzz Aldrin, and Michael Collins*)
- What was significant about the Apollo 11 spaceflight in 1969? (*During this mission, astronauts walked on the moon for the first time.*)
- During the Apollo 11 spaceflight, Michael Collins' job was to remain in the spacecraft and orbit the moon by himself while Neil Armstrong and Buzz Aldrin did what? (*Neil Armstrong and Buzz Aldrin landed the Lunar Module on the surface of the moon and became the first men ever to walk on the moon.*)
- Who was the first man ever to walk on the moon? (*Neil Armstrong, followed minutes later by Buzz Aldrin*)
- What is a shooting star? (*It's not really a star. It's a meteor, like a piece of rock from space, that burns up as it moves through the earth's atmosphere.*)
- What are the eight planets in our solar system? (*Mercury, Venus, Earth, Mars, Jupiter, Saturn, Uranus, Neptune*) (**Note:** There is an acronym with "The Solar System" page to help students learn these planets and their distance from the sun.)
- What is the largest planet in our solar system? (*Jupiter*)
- What is the smallest planet in our solar system? (*Mercury*)
- The galaxy we are in contains earth and other planets of our solar system, as well as our sun and billions of other stars. Identify the galaxy. (*Milky Way Galaxy*)

Activities, Supplies, and Preparations for this Lesson

- **Spelling Patterns of Sounds Heard in the English Language:** Display words with this week's sound on the wall display. (See pages 26-27 for details.)
- **Reproducible materials:** Prepare the reproducible materials provided with this lesson, including the "Spelling Word Riddles," the question cards, and the spelling word cards
(<u>Supplies:</u> Computer paper, light blue and yellow cardstock, magnetic tape)
- **Books and videos:** Read books and/or watch YouTube videos for kids about the Apollo 11 spaceflight, about shooting stars, about the eight planets in our solar system, about the Milky Way Galaxy, and about the International Space Station (ISS). For any books you read, identify the title, the author, and the illustrator. Also, examine the front and back cover of the book before making any predictions about the story or discussing what it is about.

- **Moon crafts:** Choose a moon craft activity for kids from online and display their creations in your classroom.
- **The Solar System:** Review "The Solar System" page with students and use the provided acronym to help them learn the planets and their distances (closest to furthest) from the sun. Please note that students can use the acronym to practice identifying the planets on their own with the free computer-based test for this lesson. (See page 13 for more information.) Display a laminated solar system poster and practice identifying the eight planets regularly throughout the remainder of the year--*Mercury, Venus, Earth, Mars, Jupiter, Saturn, Uranus, Neptune.*
  (<u>Supplies:</u> A laminated solar system poster; jelly and nutty peanut butter to eat upon bread--**BEWARE** that none of your students are allergic to peanuts.)
- **Thematic Units Game for Planets:** Reproduce the unit questions and the spinner from page 98 on cardstock and cut apart the pieces. Use wall-safe tape to attach the questions to different planets on a laminated solar system poster. Question cards from previous lessons can also be taped to the poster. As a variation or in addition to using the unit questions, you might choose to use the spelling word cards with the game. To prepare the spinner, put a brad through the end of a large paperclip and push it through the middle of the spinner. Turn the spinner over to fasten the brad to the back. A student spins the paperclip and chooses a question to answer or a spelling word to read from the planet it lands on.

  If you visit the "Free Teaching Materials" page of the Habakkuk Educational Materials website, there is also a free outer space gameboard to use with the unit questions. Click on "Board Games and More" to access the game.
  (<u>Supplies:</u> A laminated solar system poster, cardstock, a brad, a large paperclip, wall-safe tape)

- **Writing/Handwriting assignment:** Position one of the handwriting assignments from pages 94-95 under a Smart Document Camera or a similar device. Discuss the correct stroke for each letter as you copy the sentence onto your paper. Students should follow along with you, copying the letters on papers of their own. (This will help them to remain on task and to finish the handwriting assignment in unison.)
  (<u>Supplies:</u> Provide Milky Way milk chocolate minis for students to eat when discussing the Milky Way Galaxy.)

  <u>Older Students:</u> Older students can also write about the topic below. Let them copy the topic and respond to it in their writing journals. They may also want to add appropriate illustrations.

  Explain why you would or would not want to travel to space.

Context Clues

Directions: Fill in each blank with a spelling word that makes sense in the context of the sentence. (Cross out the words from the list as you use them.) Circle the letters in the key words that make the /ü/ sound heard in *moon* (*oo*, *ew*, *u*, *ui*, *ough*).

Spelling words: moon, shooting, zoom, crew, flew, view, lunar, Jupiter, Neptune, Pluto, spacesuit, through

1. Have you ever wished you could __________ through space in a rocket like some astronauts?

2. Astronauts Neil Armstrong, Buzz Aldrin, and Michael Collins had that experience as the __________ of Apollo 11.

3. It was Michael Collins' duty to remain in the spacecraft and orbit the moon by himself while Neil Armstrong and Buzz Aldrin landed the __________ Module on the surface of the moon.

4. On July 20, 1969, Neil Armstrong and Buzz Aldrin became the first men ever to walk on the __________.

5. There are videos you can __________ on the internet that show this historical event.

6. You can see Neil Armstrong taking the first step on the moon while dressed in a _______________.

7. Astronauts, like these three men who __________ to the moon, know a lot about space.

8. Have you ever seen a star shooting __________ the sky?

9. Astronauts understand that a __________ star is not really a star at all. It's a meteor (like a piece of rock from space) that burns up as it moves through the earth's atmosphere.

10. They could also tell you that there are eight planets in our solar system: Mercury, Venus, Earth, Mars, Jupiter, Saturn, Uranus, and __________.

11. They know that __________ is the largest planet in our solar system, and Mercury is the smallest.

12. __________ is actually smaller than Mercury, but Pluto is no longer classified as a planet, which is why it is not listed above. It is called a "dwarf planet."

Spelling Word Riddles

Discussion: All spelling words this week have the /ü/ sound in moon and can be spelled as ***oo*** as in *moon*, ***ew*** as in *crew*, ***u*** as in *Pluto*, ***ui*** as in *spacesuit*, or ***ough*** as in *through*. (Help students hear and pronounce the /ü/ sound by first saying the sound and then a basic word, such as *crew*.)

Directions: Reproduce the next two pages on light blue or yellow cardstock as instructed. Cut out the pieces and attach a magnetic strip to the back of each piece. Line up the yellow pieces horizontally towards the top of a magnetic board. Position the light blue strips vertically below them. (See the illustration on page 15.) Read a riddle and call on a student to identify the correct spelling word. For the first sentence, a student would identify the word *zoom* and then move the yellow magnetic strip with the letters ***oo*** between the light blue letters ***z*** and ***m*** to form the word ***zoom***.

1. In this word, the /ü/ sound is spelled *o-o*. It means "to move rapidly." (*Zoom*)
2. In this word, the /ü/ sound is spelled *u-i*. This is what an astronaut wears. (*Spacesuit*)
3. In this word, the /ü/ sound is spelled *u*. This is the largest planet in our solar system. (Jupiter)
4. In this word, the /ü/ sound is spelled *e-w*. This is the past tense of *fly*. (*Flew*)
5. In this word, the /ü/ sound is spelled *o-o*. This lights up the sky at night. (*Moon*)
6. In this word, the /ü/ sound is spelled *o-u-g-h*. This is going in one side and coming out the other. (*Through*)
7. In this word, the /ü/ sound is spelled *o-o*. A meteor can sometimes be seen doing this through the night sky. (*Shooting*)
8. In this word, the /ü/ sound is spelled *u*. Neil Armstrong and Buzz Aldrin landed the __________ Module on the surface of the moon. (*Lunar*)
9. In this word, the /ü/ sound is spelled *u*. This is a dwarf planet but is not considered a planet of our solar system. (*Pluto*)
10. In this word, the /ü/ sound is spelled *e-w*. It means the same thing as *see*. (*view*)
11. In this word, the /ü/ sound is spelled *u*. This planet begins with the letter *N*. (*Neptune*)
12. In this word, the /ü/ sound is spelled *e-w*. It begins with the /k/ sound and rhymes with *flew*. (*Crew*)

Directions: Use light blue cardstock to reproduce this page. Cut out the pieces and attach magnetic strips to the back of each piece. (See the previous page for further instructions.)

m	n	sh	ting
z	m	cr	
fl	v	w	
l	nar	J	piter
Nept	ne	Pl	to
spaces	t	thr	

Directions: Use yellow cardstock to reproduce this page. Cut out the pieces and attach magnetic strips to the backs.

oo	oo	oo	oo	oo	
ew	ew	ew	ew	ew	
u	u	u	u	ui	ough

Spelling Patterns of Sounds Heard in Our Spelling Words

Directions: Fill in the missing letters that make the /ü/ sound in *moon*.

- Use a **blue** colored pencil for ***oo***.
- Use a **red** colored pencil for ***ew***.
- Use a **green** colored pencil for ***u***.
- Use a **purple** colored pencil for ***ui***.
- Use a **brown** colored pencil for ***ough***.

m _ _ n	f l _ _	N e p t _ n e
s h _ _ t i n g	v i _ _	P l _ t o
z _ _ m	l _ n a r	s p a c e s _ _ t
c r _ _	J _ p i t e r	t h r _ _ _ _

Directions: Record spelling words whose /ü/ sound is spelled *oo* beside the **"oo"** heading, words whose /ü/ sound is spelled *ew* beside the **"ew"** heading, and so forth.

oo ____

ew ____

u ____

ui ____

ough ____

Spelling Words (/ü/ sound in *moon*)

G	C	N	V	B	T	H	R	O	U	G	H
N	V	I	E	W	K	S	U	T	F	B	X
E	U	O	D	M	A	P	L	U	T	O	V
P	T	F	M	H	T	Y	V	D	X	H	S
T	P	L	C	C	R	E	W	J	Z	D	H
U	I	E	F	J	D	O	A	L	O	N	O
N	A	W	Q	V	K	M	W	U	O	S	O
E	F	N	M	O	O	N	P	N	M	T	T
G	S	L	W	E	G	H	B	A	R	W	I
J	U	P	I	T	E	R	P	R	I	Q	N
I	R	Q	E	I	P	X	Y	V	F	M	G
S	P	A	C	E	S	U	I	T	P	J	X

Space Words

1. moon
2. shooting
3. zoom
4. crew
5. flew
6. view
7. lunar
8. Jupiter
9. Neptune
10. Pluto
11. spacesuit
12. through

Homophones

flew	flu

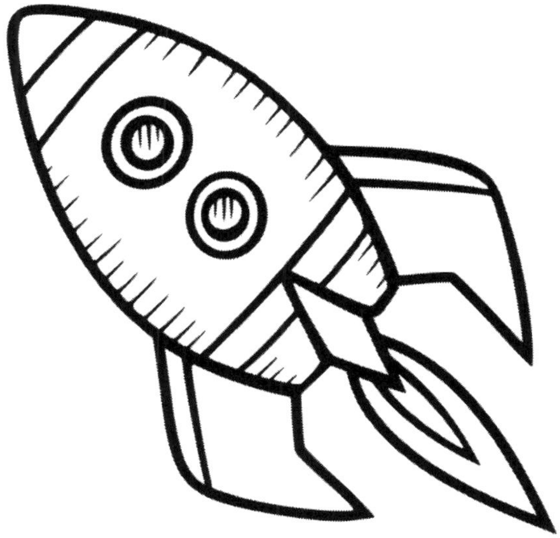

The rocket **flew** through space.

Flew is the past tense of "fly."

I've been sick with the **flu** all weekend!

The flu is an illness. It is short for "influenza."

I **flew** home early because my child has the **flu**.

Underline the word that is the past tense of *fly*.
Circle the word that refers to an illness.

Homophones

threw

The pitcher **threw** the ball.

Threw is the past tense of "throw."

through

The ball went **through** the neighbor's window.

Through is going in one side and coming out the other side.

The pitcher **threw** the ball right **through** a neighbor's window!

Underline the word that is the past tense of *throw*.
Circle the word that means to go in one side and come out the other.

Compound Words

Directions: When two or more root words are used together to form one word, it is called a **compound word** (spacecraft). Form four compound words by adding a spelling word to the words *space*, *out*, *full*, and *stars*.

Word list: moon, shooting, suits, through

space_________ _________out full _________

_________ stars

Directions: Fill in each blank with one of the compound words that makes sense in the context of the sentence.

The astronauts put on their ________________ and boarded the spacecraft. Then they traveled ________________ space to get to the International Space Station. From the station, they saw some ________________ and also had an awesome view of the ________________.

Weekly Oral Language

1. Correct the asking sentence.

   what are the eight planets in our solar system

2. Correct the telling sentence.

   the eight planets are Mercury, Venus, Earth, Mars, Jupiter, Saturn, Uranus, and Neptune

3. What is the subject of the telling sentence?

   ____

4. Read the *telling sentence* again and identify the verb.

   ____

5. Write your own sentence about this week's theme.

   ____

Instruction: *Teaching Children about Capitalization, Punctuation, Apostrophe s, and Contractions (Language Book 2)* by Habakkuk Educational Materials can be purchased separately to give students the knowledge they need to complete this worksheet. After students have completed the worksheet and turned it in, complete a copy of it on your classroom screen.

Parts of Speech

Directions: Identify what parts of speech are used in the sentence below and write them in the spaces provided. Then identify the subject (naming part) and predicate (action part) of the sentence and write them in the spaces provided.

Parts of speech: Noun marker, noun, pronoun, verb, helping verb, adjective, adverb, preposition, conjunction, interjection

Jupiter	is	a	huge	planet.
________	________	________	________	________

Subject (Naming Part)

____

Predicate (Action Part)

____

Instruction: *Teaching Children about Parts of Speech (Language Book 3)* by Habakkuk Educational Materials can be purchased separately to help students quickly learn the parts of speech. After students have completed this worksheet and turned it in, complete a copy of it on your classroom screen.

Neil Armstrong was the first man ever to step on the moon. A few minutes later, his friend Buzz Aldrin also stepped out of the Lunar Module and walked upon the moon's surface. While there, they put an American flag in the ground of the moon.

Directions: Read the paragraph above. Then copy the sentence below on the lines at the bottom of the page. You can also color the picture.

On July 20, 1969, Neil Armstrong became the first man ever to walk on the moon.

The illustration on this page is by Lena London and is licensed under CC BY-SA 4.0.
http://www.supercoloring.com/coloring-pages/first-man-on-the-moon

The Milky Way Galaxy

The galaxy we are in is called the Milky Way. The Milky Way contains earth and other planets of our solar system, as well as our sun and billions of other stars.

Directions: Read the paragraph above. Then copy the sentence below on the lines at the bottom of the page. You can also color the picture.

We are in the Milky Way Galaxy.

The Solar System

Directions: Use the below acronym to help students learn the names of the planets and their distance (closest to furthest) from the sun. Point out the comets and asteroids, as well as Pluto, which is no longer considered a planet. (You might even want to let your students taste jelly and nutty peanut butter upon bread, providing that none of your students have nut allergies.) When students are familiar with the acronym, they can practice identifying the planets on their own with the free computer-based test for this lesson. (See page 13 for more information.)

*Mrs. Venus eats Mark's jelly spread upon nutty **p**eanut butter.*

Mrs. stands for *Mercury*, ***Venus*** is unchanged, ***eats*** stands for *Earth*, ***Mark's*** for *Mars*, ***Jelly*** for *Jupiter*, ***spread*** for *Saturn*, ***upon*** for *Uranus*, and ***nutty*** for *Neptune*. In addition, ***peanut butter*** stands for *Pluto*, although it is no longer considered a planet. (There are two planets that start with *M*. Tell students that the name *Mark* is representing *Mars*, as it almost has the planet spelled within it.)

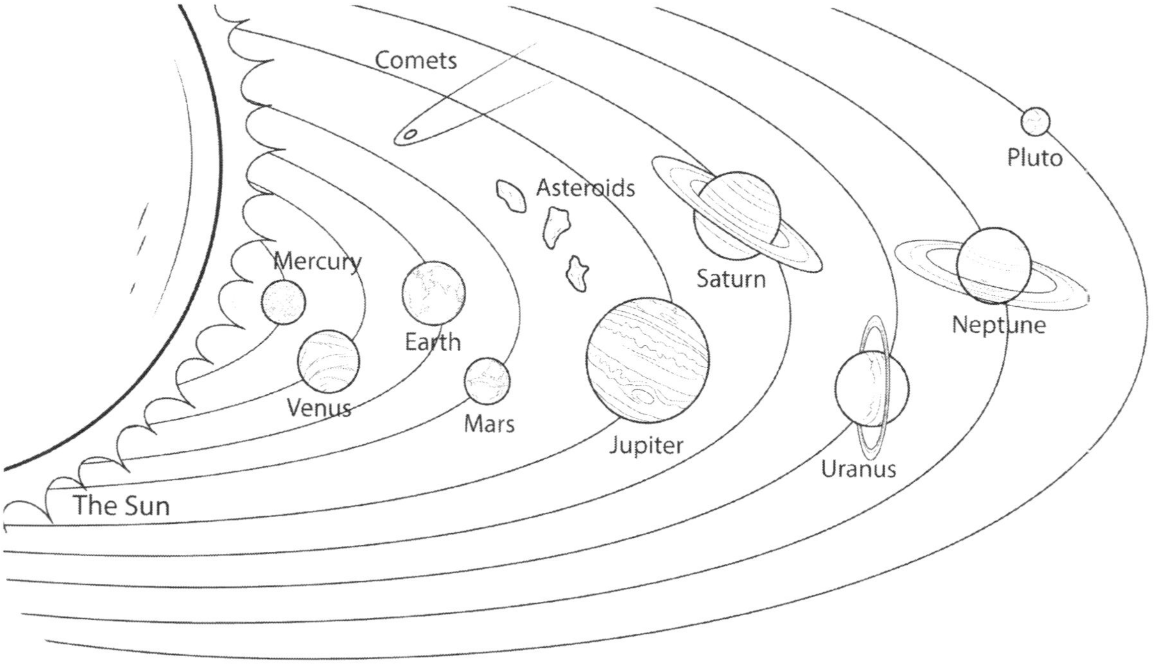

Name: ______________________________

Addition and Subtraction Facts

Directions

1. Count the rockets above the first space and write the addend.
2. Count the rockets above the second space and write the addend.
3. Count all the rockets and write the sum after the equal sign.

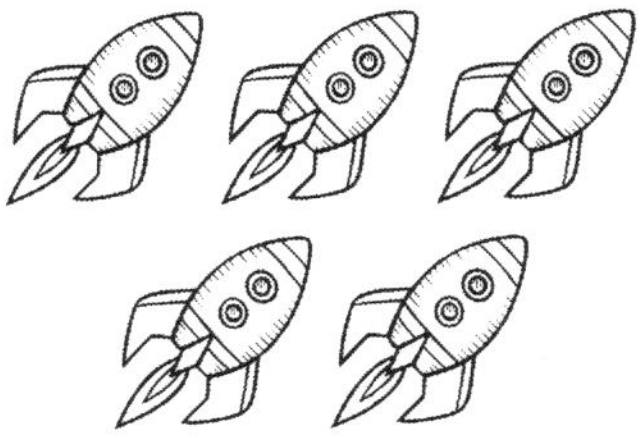

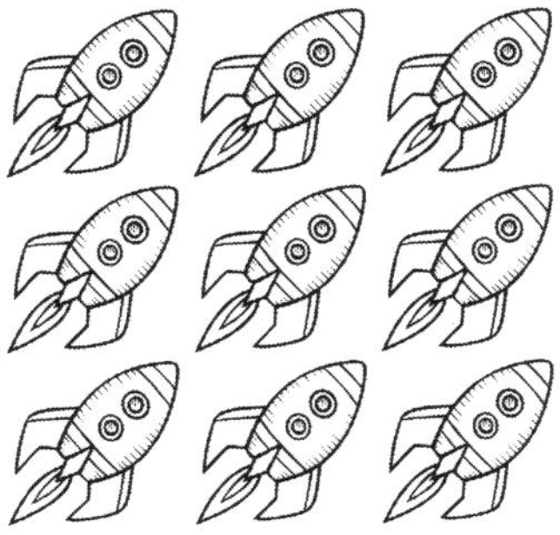

_________ + _________ = _________

Directions

1. Count the stars above the first space and write the number.
2. Cross out 2 stars.
3. Count the number of stars left and write the answer after the equal sign.

_________ − ___2___ = _________

QUESTIONS FOR THE THEMATIC UNITS GAME

Please refer to pages 28-29 of this book for instructions on how to use the unit questions.

Who was the crew of the Apollo 11 spaceflight in 1969? *Lesson 4 (Science Themes)*	What was significant about the Apollo 11 spaceflight in 1969? *Lesson 4 (Science Themes)*
During the Apollo 11 spaceflight, Michael Collins' job was to remain in the spacecraft and orbit the moon by himself while Neil Armstrong and Buzz Aldrin did what? *Lesson 4 (Science Themes)*	Who was the first man ever to walk on the moon? *Lesson 4 (Science Themes)*
What is a shooting star? *Lesson 4 (Science Themes)*	What are the eight planets in our solar system? *Lesson 4 (Science Themes)*
What is the largest planet in our solar system? *Lesson 4 (Science Themes)*	What is the smallest planet in our solar system? *Lesson 4 (Science Themes)*
The galaxy we are in contains earth and other planets of our solar system, as well as our sun and billions of other stars. Identify the galaxy. *Lesson 4 (Science Themes)*	

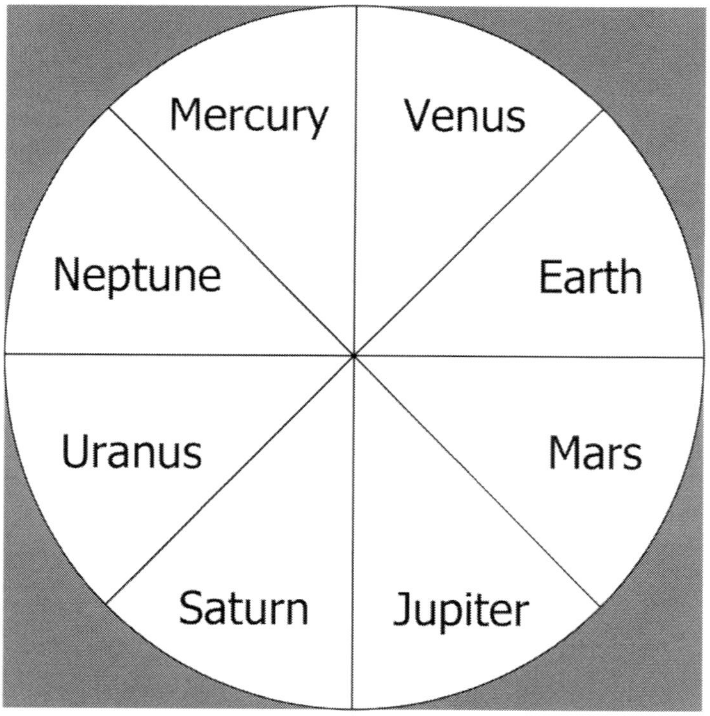

Note: The unit questions and the spinner can be used for the thematic units game described on page 81.

LESSON 4 TEST (Space)

1. Who was the crew of the Apollo 11 spaceflight in 1969?

   ○ Jim Lovell, Fred Haise, and Jack Swigert

   ○ Charles "Pete" Conrad Jr., Richard F. Gordon Jr., and Alan L. Bean

   ○ Neil Armstrong, Buzz Aldrin, and Michael Collins

2. What was significant about the Apollo 11 spaceflight in 1969?

   ○ Astronauts walked on the moon for the first time.

   ○ It was the first mission into space.

   ○ The astronauts almost didn't make it back to earth.

3. During the Apollo 11 spaceflight, Michael Collins' job was to remain in the spacecraft and orbit the moon by himself while Neil Armstrong and Buzz Aldrin did what?

   ○ Landed the Lunar Module on the surface of the moon

   ○ Became the first men ever to walk on the moon

   ○ Both of the above answers are true.

4. Who was the first man ever to walk on the moon?

   ○ Charles "Pete" Conrad Jr.

   ○ Neil Armstrong, followed minutes later by Buzz Aldrin

   ○ Jim Lovell

5. What is a shooting star?

   ○ A star that is falling from the sky

   ○ A meteor, like a piece of rock from space, that burns up as it moves through the earth's atmosphere

   ○ Both of the above answers are true.

6. What are the eight planets in our solar system? (Mark two answers.)

- ◯ Sun, Pluto, Comet, Asteroid
- ◯ Mercury, Venus, Earth, Mars
- ◯ Jupiter, Saturn, Uranus, Neptune

7. What is the largest planet in our solar system?

- ◯ Earth
- ◯ Mercury
- ◯ Jupiter

8. What is the smallest planet in our solar system?

- ◯ Earth
- ◯ Mercury
- ◯ Jupiter

9. The galaxy we are in contains earth and other planets of our solar system, as well as our sun and billions of other stars. Identify the galaxy.

- ◯ Milky Way Galaxy
- ◯ Sombrero Galaxy
- ◯ Whirlpool Galaxy

10. Use words from the word list to identify the asteroids, the comets, the eight planets, and Pluto.

Word list: Asteroids, comet, Earth, Jupiter, Mars, Mercury, Neptune, Pluto, Saturn, Sun, Uranus, Venus

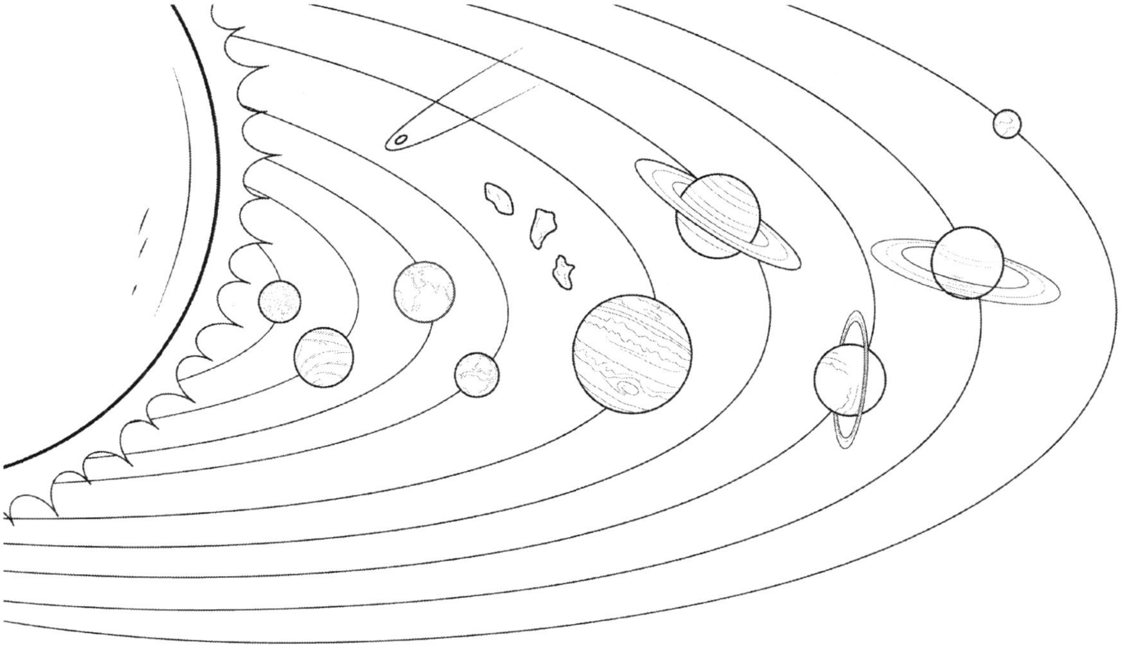

Note: These cards can be used for the spelling word games listed on pages 21-22.

m<u>oo</u>n	sh<u>oo</u>ting
z<u>oo</u>m	cr<u>ew</u>
fl<u>ew</u>	vi<u>ew</u>

Note: These cards can be used for the spelling word games listed on pages 21-22.

lunar	Jupiter
Neptune	Pluto
spacesuit	through

Lesson 5

Theme: The Wild West
Sound: diphthong /ou/
Spelling Patterns: *ou, ow*

Spelling Words	Alphabetized
1. cows	________________
2. cowboy	________________
3. chow	________________
4. howdy	________________
5. town	________________
6. brown	________________
7. how	________________
8. howl	________________
9. sounds	________________
10. clouds	________________
11. ground	________________
12. outside	________________

Inflectional endings: howls, howling, howled

Bonus compound word: cowgirl

Extra challenge word: outlaw

Context Clues

Directions: Fill in each blank with a spelling word that makes sense in the context of the sentence. (Cross out the words from the list as you use them.) Circle the letters in the key words that make the /ou/ sound (*ow*, *ou*).

Spelling words: cows, cowboy, chow, howdy, town, brown, how, howl, sounds, clouds, ground, outside

1. ____________, Partner!

2. I'm a ____________ of the Wild West!

3. I usually live in ____________, but right now I'm on a cattle drive with some other cowboys.

4. Our job is to move a bunch of ____________ to another state.

5. I have to show the young cowboys ____________ it's done.

6. While we're away from home, we sleep ____________ with the cattle.

7. A blanket on the ____________ is our bed.

8. Sometimes we hear strange ____________.

9. When wolves ____________ at night, we sing to the cattle to calm them down.

10. The first thing I saw when I woke up this morning were the ____________ in the sky.

11. Then I saw some ____________ cows grazing.

12. We're going to eat some ____________ and then get back to moving the cows.

Spelling Word Riddles

Discussion: All spelling words this week have the /ou/ sound and can be spelled as ***ow*** as in *cow* or ***ou*** as in *ground*. (Help students hear and pronounce the /ou/ sound by first saying the sound and then a basic word, such as *out*.)

Directions: Reproduce the next two pages on light blue or yellow cardstock as instructed. Cut out the pieces and attach a magnetic strip to the back of each piece. Line up the yellow pieces horizontally towards the top of a magnetic board. Position the light blue strips vertically below them. (See the illustration on page 15.) Read a riddle and call on a student to identify the correct spelling word. For the first sentence, a student would identify the word *brown* and then move the yellow magnetic strip with the letters ***ow*** between the light blue letters ***br*** and ***n*** to form the word ***brown***.

1. In this word, the /ou/ sound is spelled *o-w*. This is the color of some cows. (*Brown*)
2. In this word, the /ou/ sound is spelled *o-w*. This is how cowboys sometimes said hello. (*Howdy*)
3. In this word, the /ou/ sound is spelled *o-u*. You will probably see these if you look up at the sky. (*Clouds*)
4. In this word, the /ou/ sound is spelled *o-w*. You can get milk from these animals. (*Cows*)
5. In this word, the /ou/ sound is spelled *o-w*. This is what cowboys called their food. (*Chow*)
6. In this word, the /ou/ sound is spelled *o-u*. Cowboys slept on this while they were on their cow drives. (*Ground*)
7. In this word, the /ou/ sound is spelled *o-w*. It begins with *h* and rhymes with *cow*. (*How*)
8. In this word, the /ou/ sound is spelled *o-u*. It means the opposite of *inside*. (*Outside*)
9. In this word, the /ou/ sound is spelled *o-w*. You might hear a wolf do this. (*Howl*)
10. In this word, the /ou/ sound is spelled *o-u*. This is something you can hear with your ears. (*Sounds*)
11. In this word, the /ou/ sound is spelled *o-w*. This is a place that usually has a lot of stores and buildings. (*Town*)
12. In this word, the /ou/ sound is spelled *o-w*. In the Wild West, this man wore a hat, rode a horse, and worked with cows. (*Cowboy*)

Directions: Use light blue cardstock to reproduce this page. Cut out the pieces and attach magnetic strips to the back of each piece. (See the previous page for further instructions.)

c	s	c	boy
ch	h	dy	
t	n	br	n
h	h	l	
s	nds	cl	ds
gr	nd	tside	

Directions: Use yellow cardstock to reproduce this page. Cut out the pieces and attach magnetic strips to the backs.

ow	ow	ow	ow
ow	ow	ow	ow

ou	ou	ou	ou	ou

Spelling Patterns of Sounds Heard in Our Spelling Words

Directions: Fill in the missing letters that make the /ou/ sound.

- Use a **blue** colored pencil for ***ow***.
- Use a **red** colored pencil for ***ou***.

c _ _ s	t _ _ n	s _ _ n d s
c _ _ b o y	b r _ _ n	c l _ _ d s
c h _ _	h _ _	g r _ _ n d
h _ _ d y	h _ _ l	_ _ t s i d e

Directions: Record spelling words whose /ou/ sound is spelled *ow* under the **"ow"** headings and words whose /ou/ sound is spelled *ou* under the **"ou"** heading.

ow	ow	ou

Spelling Words (/ou/ sound)

H	C	G	V	B	O	U	T	S	I	D	E
O	S	R	I	L	K	S	U	T	F	B	X
W	U	O	D	S	O	U	N	D	S	J	C
C	T	U	M	H	T	Y	V	D	X	H	L
K	P	N	C	T	N	R	H	J	B	D	O
H	I	D	F	O	D	O	O	L	R	N	U
O	D	A	Q	W	K	M	W	C	O	S	D
W	F	N	Z	N	F	H	L	K	W	T	S
D	S	L	W	E	G	H	B	C	N	W	O
Y	G	C	H	O	W	E	P	H	I	Q	C
I	R	Q	E	I	P	C	O	W	B	O	Y
Q	C	O	W	S	Z	U	I	J	P	J	X

The Wild West

1. cows
2. cowboy
3. chow
4. howdy
5. town
6. brown
7. how
8. howl
9. sounds
10. clouds
11. ground
12. outside

Compound Words

Directions: When two or more root words are used together to form one word, it is called a **compound word** (cowgirl). Form three compound words by adding a word from the word list to the words *boy*, *side*, and *hat*.

Word list: cow, cowboy, out

_________boy _________side _________ hat

Directions: Fill in each blank with one of the compound words that makes sense in the context of the sentence.

I'm a __________________ on a cattle drive. At night, I sleep __________________ with the cows. During the day, I wear a __________________ on my head to keep the sun out of my eyes.

Weekly Oral Language

1. Correct the asking sentence.

   where are the cows

2. Correct the telling sentence.

   the cows are grazing in the field

3. What is the subject of the telling sentence?

   ____

4. Read the *telling sentence* again and identify the helping verb and the verb (action word).

   Helping Verb: ______________________

   Verb: ______________________________

5. Write your own sentence about this week's theme.

   ____

Instruction: *Teaching Children about Capitalization, Punctuation, Apostrophe s, and Contractions (Language Book 2)* by Habakkuk Educational Materials can be purchased separately to give students the knowledge they need to complete this worksheet. After students have completed the worksheet and turned it in, complete a copy of it on your classroom screen.

Parts of Speech

Directions: Identify what parts of speech are used in the sentence below and write them in the spaces provided. Then identify the subject (naming part) and predicate (action part) of the sentence and write them in the spaces provided.

Parts of speech: Noun marker, noun, pronoun, verb, helping verb, adjective, adverb, preposition, conjunction, interjection

The	**cowboy**	**sang**	**a**	**song.**
__________	__________	__________	__________	__________

Subject (Naming Part)

____

Predicate (Action Part)

____

Instruction: *Teaching Children about Parts of Speech (Language Book 3)* by Habakkuk Educational Materials can be purchased separately to help students quickly learn the parts of speech. After students have completed this worksheet and turned it in, complete a copy of it on your classroom screen.

Diphthongs *ou, ow, oi,* and *oy*

Spelling with *ou* and *ow*: The diphthongs *ou* as in *out* and *ow* as in *cow* make the same sound, but

- ***ou*** is usually found at the beginning of a word or syllable or in the middle of it (out, shout), whereas
- ***ow*** is usually found at the end of a word (cow).

Spelling with *oi* and *oy*: The diphthongs *oi* as in *oil* and *oy* as in *boy* make the same sound, but

- ***oi*** is usually found at the beginning of a word or syllable or in the middle of it (oil, coin), whereas
- ***oy*** is usually found at the end of a word (boy).

Out, Sh**ou**t, c**ow**	**Oi**l, C**oi**n, B**oy**
ou_____	oi_____
____ou____	____oi____
_____ow	_____oy

____

____

Diphthong: Two vowel sounds that come together so fast that they are considered one syllable

Examples: ***ou*** (ouch, our, found, house); ***ow*** (how, now); ***oi*** (oink, join, point); ***oy*** (joy, toy)

Name: ______________________________

Addition and Subtraction Facts

Directions

1. Count the cows above the first space and write the addend.
2. Count the cows above the second space and write the addend.
3. Count all the cows and write the sum after the equal sign.

______ + ______ = ______

Directions

1. Count the cowboy hats above the first space and write the number.
2. Cross out 3 cowboy hats.
3. Count the number of cowboy hats left and write the answer after the equal sign.

______ − 3 = ______

Note: These cards can be used for the spelling word games listed on pages 21-22.

c<u>ow</u>s	c<u>ow</u>boy
ch<u>ow</u>	h<u>ow</u>dy
t<u>ow</u>n	br<u>ow</u>n

Note: These cards can be used for the spelling word games listed on pages 21-22.

h<u>ow</u>	h<u>ow</u>l
s<u>ou</u>nds	cl<u>ou</u>ds
gr<u>ou</u>nd	<u>ou</u>tside

Lesson 6

Theme: Real Bears and Teddy Bears
Sound: /ĕər/ in *bear*
Spelling Patterns: *ear, air, are, arr, err, ere*

Spelling Words	Alphabetized
1. bear	______________
2. wear	______________
3. tear	______________
4. chair	______________
5. hair	______________
6. care	______________
7. scare	______________
8. share	______________
9. carry	______________
10. berry	______________
11. berries	______________
12. where	______________

Bonus compound words: blackberry, blueberry, redberry, everywhere, somewhere, teddy bear

Thematic Activities for Lesson 6
(Real Bears and Teddy Bears)

Thematic objectives: Students will learn about different kinds of bears (grizzly bears, panda bears, polar bears, and even teddy bears).

Key Questions for Discussion

- What kind of bear has white fur, likes the cold, and lives on land that surrounds the Arctic Ocean in places like Canada, Alaska, and Russia? (*polar bear*)
- What kind of bear is black and white and lives in China? (*panda bear*)

Activities, Supplies, and Preparations for this Lesson

- **Teddy Bears' Picnic:** Send a note to parents, asking them to bring their child's favorite teddy bear to school along with a sack lunch and drink. Try to have a picnic basket and picnic blanket available as well for a special teddy bears' picnic at the park or on your school's playground.
- **Spelling Patterns of Sounds Heard in the English Language:** Display words with this week's sound on the wall display. (See pages 26-27 for details.)
- **Reproducible materials:** Prepare the reproducible materials provided with this lesson, including the "Spelling Word Riddles," the question cards, and the spelling word cards
  (Supplies: Computer paper, light blue and yellow cardstock, magnetic tape)

- **Books and videos:** Read books and/or watch YouTube videos about teddy bears and different kinds of real bears (e.g., panda bears, polar bears, grizzly bears). For any books you read, identify the title, the author, and the illustrator. Also, examine the front and back cover of the book before making any predictions about the story or discussing what it is about.
- **Thematic Units Game for Bears:** Provide pictures of several different kinds of bears, such as grizzly bears, panda bears, polar bears, and teddy bears. Tape each picture to a lunch bag. Reproduce the unit questions from page 135 on cardstock and distribute them between the bags. Question cards from previous lessons should also be added to the bags. Record the names of the bears in the spaces of a blank spinner. A student spins the spinner, reads the name of the bear that the spinner lands on, and then draws a question card from the lunch bag with the bear's picture. If he answers the question correctly, his team receives one point.

- **Writing/Handwriting assignment:** Position the handwriting assignment on page 132 (i.e., the page with the teddy bear) under a Smart Document Camera or a similar device. Discuss the correct stroke for each letter as you copy the sentence onto your paper. Students should follow along with you, copying the letters on papers of their own. (This will help them to remain on task and to finish the handwriting assignment in unison.)

  <u>Older Students:</u> Older students can also complete the "What is your favorite kind of bear and why?" worksheet included with this lesson.

Context Clues

Directions: Fill in each blank with a spelling word that makes sense in the context of the sentence. (Cross out the words from the list as you use them.) Circle the letters in the key words that make the /ĕər/ sound heard in *b<u>ear</u>* (*ear, air, are, arr, err, ere*).

Spelling words: bear, wear, tear, chair, hair, care, scare, share, carry, berry, berries, where

1. I have a teddy ____________.
2. I even have different clothes for him to ____________.
3. I ____________ him around almost everywhere.
4. Sometimes, though, I ____________ my bear with my baby sister.
5. When we're not holding him, he has his very own ____________ to sit in.
6. That way I always know ____________ he is.
7. I ____________ about my teddy bear.
8. My dog once tried to ____________ him up, but I saved my bear.
9. Real bears can ____________ people, but they are awesome too.
10. Their ____________ is called *fur*.
11. They like to eat all different kinds of ____________.
12. What is your favorite kind of ____________ to eat?

Spelling Word Riddles

Discussion: All spelling words this week have the /ĕər/ sound and can be spelled as ***ear*** as in *bear*, ***air*** as in *hair*, ***are*** as in *care*, ***arr*** as in *carry*, ***err*** as in *berry*, or ***ere*** as in *where*. (Help students hear and pronounce the /ĕər/ sound by first saying the sound and then a basic word, such as *bear*.)

Directions: Reproduce the next page on light blue and yellow cardstock. Cut out the pieces and attach a magnetic strip to the back of each piece. Line up the yellow pieces horizontally towards the top of a magnetic board. Position the light blue strips vertically below them. (See the illustration on page 15.) Read a riddle and call on a student to identify the correct spelling word. For the first sentence, a student would identify the word *chair* and then move the yellow magnetic strip with the letters ***air*** to the right of the light blue magnetic strip having the letters ***ch*** to form the word ***chair***.

1. In this word, the /ĕər/ sound is spelled *a-i-r*. This is something you sit on. (*Chair*)
2. In this word, the /ĕər/ sound is spelled *e-a-r*. This is what you do with a hat. (*Wear*)
3. In this word, the /ĕər/ sound is spelled *e-r-e*. It is often used to ask a question. (*Where*)
4. In this word, the /ĕər/ sound is spelled *e-a-r*. This means "to rip." (*Tear*)
5. In this word, the /ĕər/ sound is spelled *a-r-e*. Bears sometimes do this to people. (*Scare*)
6. In this word, the /ĕər/ sound is spelled *e-r-r*. I pulled one of these from a bush and ate it. (*Berry*)
7. In this word, the /ĕər/ sound is spelled *e-r-r*. This is the plural of *berry*. (*Berries*)
8. In this word, the /ĕər/ sound is spelled *a-r-e*. When you help someone, it shows them that you _____. (*care*)
9. In this word, the /ĕər/ sound is spelled *a-i-r*. This is white on a polar bear. (*Hair*)
10. In this word, the /ĕər/ sound is spelled *e-a-r*. This is an animal. (*Bear*)
11. In this word, the /ĕər/ sound is spelled *a-r-e*. It's nice to do this with your toys. (*Share*)
12. In this word, the /ĕər/ sound is spelled *a-r-r*. How would you move a teddy bear from one room to another? (*Carry*)

b	w	t	ch
h	c	sc	sh
c	y	b	y
b	ies	wh	

Directions: Reproduce the cards above on light blue cardstock and the cards below on yellow cardstock. Cut out the pieces and attach magnetic strips to the backs. (See the previous page for further instructions.)

air	air	err	err	arr
ear	ear	ear	are	
are	are	ere	ere	

Spelling Patterns of Sounds Heard in Our Spelling Words

Directions: Fill in the missing letters that make the /ĕər/ sound in *b<u>ear</u>*.

- Use a **blue** colored pencil for ***ear***.
- Use a **red** colored pencil for ***air***.
- Use a **green** colored pencil for ***are***.
- Use a **purple** colored pencil for ***arr***.
- Use a **brown** colored pencil for ***err***.
- Use an **orange** colored pencil for ***ere***.

b _ _ _	h _ _ _	c _ _ _ y
w _ _ _	c _ _ _	b _ _ _ y
t _ _ _	s c _ _ _	b _ _ _ i e s
c h _ _ _	s h _ _ _	w h _ _ _

Directions: Record spelling words whose /ĕər/ sound is spelled *ear* beside the **"ear"** heading, words whose /ĕər/ sound is spelled *air* beside the **"air"** heading, and so forth.

ear ______________________________

air ______________________________

are ______________________________

arr ______________________________

err ______________________________

ere ______________________________

Spelling Words (/ĕər/ sound in *bear*)

K	C	N	V	B	W	A	H	A	I	R	O
C	A	R	E	L	H	S	U	T	F	B	X
G	U	O	D	M	E	R	B	E	R	R	Y
C	T	B	M	H	R	Y	V	D	X	H	Z
T	E	A	R	H	E	D	R	J	Z	D	B
B	I	N	F	C	D	S	A	L	M	N	E
B	D	A	Q	A	K	C	W	C	W	S	R
E	F	N	Z	R	F	A	P	K	E	T	R
A	S	L	W	R	G	R	B	C	A	W	I
R	G	Y	Y	Y	M	E	P	H	R	Q	E
I	R	Q	E	I	P	X	Y	V	F	M	S
C	H	A	I	R	Z	S	H	A	R	E	X

Real and Teddy Bears

1. bear
2. wear
3. tear
4. chair
5. hair
6. care
7. scare
8. share
9. carry
10. berry
11. berries
12. where

Homophones

bear

The **bear** has fun exploring the woods.

A bear is an animal.

bare

The tree is completely **bare**.

If a tree is bare, the leaves that are supposed to cover it are gone.

The **bear** approached a bush that was completely **bare**.

Underline the word that refers to an animal.
Circle the word that means that something is without its covering.

Homophones

hair

hare

The girl has curly **hair**.

The **hare** looks scared.

Hair is what grows on your head.

A hare is an animal that runs fast and that looks like a big rabbit.

This **hare** has **hair** that turns white in the winter.

Underline the word that refers to what grows on your head.
Circle the word that refers to an animal that looks like a big rabbit.

wear

I will **wear** the green shirt.

Wear means "to have on the body."

where

Where are you hiding?

Where is a question word.

Where are you going to **wear** that outfit?

Underline the word that means "to have on the body."
Circle the word that is a question word.

Compound Words

Directions: When two or more root words are used together to form one word, it is called a **compound word** (black<u>berry</u>). Form four compound words by adding a spelling word to the words *blue*, *every*, *some*, and *teddy*.

Spelling words: bear, berry, where

blue_____ every_____ some_____ teddy _____

Directions: Fill in each blank with one of the compound words that makes sense in the context of the sentence.

I have a stuffed ________________. I put it ________________, but I can't remember where. I looked ________________ for it. Then I remembered that I sat it down outside by the ________________ bush so that I could pick some of the berries to eat.

Weekly Oral Language

1. Correct the asking sentence.

   what do bears eat

2. Correct the telling sentence.

   bears eat plants, berries, insects, and fish

3. What is the subject of the telling sentence?

   ____

4. Read the *telling sentence* again and identify the verb (action word).

   ____

5. Write your own sentence about this week's theme.

   ____

Instruction: *Teaching Children about Capitalization, Punctuation, Apostrophe s, and Contractions (Language Book 2)* by Habakkuk Educational Materials can be purchased separately to give students the knowledge they need to complete this worksheet. After students have completed the worksheet and turned it in, complete a copy of it on your classroom screen.

Parts of Speech

Directions: Identify what parts of speech are used in the sentence below and write them in the spaces provided. Then identify the subject (naming part) and predicate (action part) of the sentence and write them in the spaces provided.

Parts of speech: Noun marker, noun, pronoun, verb, helping verb, adjective, adverb, preposition, conjunction, interjection

The	**bear**	**is**	**eating**	**berries.**
__________	__________	__________	__________	__________

Subject (Naming Part)

____

Predicate (Action Part)

____

Instruction: *Teaching Children about Parts of Speech (Language Book 3)* by Habakkuk Educational Materials can be purchased separately to help students quickly learn the parts of speech. After students have completed this worksheet and turned it in, complete a copy of it on your classroom screen.

Handwriting directions: Color the picture. Then copy the sentence below on the lines at the bottom of the page.

The teddy bear is on a picnic.

What is your favorite kind of bear and why?

____

____

____

____

Panda Bear

Panda bears live in China.

Polar Bear

Polar bears like the cold and live on land that surrounds the Arctic Ocean in places like Canada, Alaska, and Russia.

Grizzly Bear

Grizzly bears live in North America, Europe, and Asia

Teddy Bear

A teddy bear is a toy.

Name: ______________________________

Addition and Subtraction Facts

Directions

1. Count the bears above the first space and write the addend.
2. Count the bears above the second space and write the addend.
3. Count all the bears and write the sum after the equal sign.

_________ + _________ = _________

Directions

1. Count the bears above the first space and write the number.
2. Cross out 7 bears.
3. Count the number of bears left and write the answer after the equal sign.

_________ − 7 = _________

QUESTIONS FOR THE THEMATIC UNITS GAME

Please refer to pages 28-29 of this book for instructions on how to use the unit questions.

What kind of bear has white fur, likes the cold, and lives on land that surrounds the Arctic Ocean in places like Canada, Alaska, and Russia? *Lesson 6 (Science Themes)*	What kind of bear is black and white and lives in China? *Lesson 6 (Science Themes)*

LESSON 6 TEST (Real Bears and Teddy Bears)

1. What kind of bear has white fur, likes the cold, and lives on land that surrounds the Arctic Ocean in places like Canada, Alaska, and Russia?

   ○ Grizzly bear

   ○ Panda bear

   ○ Polar bear

2. What kind of bear is black and white and lives in China?

   ○ Grizzly bear

   ○ Panda bear

   ○ Polar bear

Note: These cards can be used for the spelling word games listed on pages 21-22.

b<u>ear</u>	w<u>ear</u>
t<u>ear</u>	ch<u>air</u>
h<u>air</u>	c<u>are</u>

Note: These cards can be used for the spelling word games listed on pages 21-22.

scare	share
carry	berry
berries	where

Lesson 7

Theme: Fire Safety
(Community Helpers: Firemen)
Sound: /f/
Spelling Patterns: *f*, *fe*

Spelling Words	Alphabetized
1. fire	______________
2. force	______________
3. fuel	______________
4. face	______________
5. fury	______________
6. forest	______________
7. fight	______________
8. inferno	______________
9. floor	______________
10. flare	______________
11. flames	______________
12. safety	______________

Inflectional endings: fuels, fueling, fueled

Bonus compound words: firefighter, firehose, fireman, firemen, firetruck, fire station, forest fire

Extra challenge word: fierce

Thematic Activities for Lesson 7
(Fire Safety)

Thematic objectives: Students will learn about fire safety, including how to avoid starting a fire, how to prepare for the event of a fire, what to do if their clothes ever catch fire, what to do if they burned their skin, how to escape a smoke-filled house, and what number to dial in an emergency. They will also understand that fire needs oxygen to keep burning.

Key Questions for Discussion

- What is the emergency number to dial if you ever need the police, an ambulance, or a fireman? (*911 in the United States and Canada, 999 in the United Kingdom, and 000 in Australia*)
- What does fire need to keep burning? (*oxygen, or air*)
- What would you do if your clothes caught on fire? (*Stop, drop, and roll—Cover your face and roll over and over again until the fire is out. You smother the fire when you stop, drop, and roll because fire needs oxygen to keep burning. Clothes can fuel a fire in the same way wood does.*)
- What would you do if you were making something to eat and burned your skin? (*If it's not a serious burn, put the burn in cool water, not ice water. If it is serious, you will also need to call the emergency number.*)
- What would you do if the smoke alarm went off while you were home alone? (*First get out of the house. Then call the emergency number from a cell or from your neighbor's home. If there is smoke in the house, crawl under the smoke to get to a door or window and/or use your shirt or something else to cover your nose and your mouth so that you don't inhale the smoke. Also, if you are in your bedroom or another room with the door closed, feel the door to make sure it is not hot before opening it. If it is hot or if the doorknob feels hot, there is fire on the other side, and you need to crawl out of the window instead.*)
- What precautions can you take to help protect you, your family, and your home from fires. (*Check the smoke alarms monthly and replace the batteries yearly. Practice crawling out of your bedroom window.*)
- What are some things a child should never do because of the fire danger? (*Never stick anything into a light socket. Do not play with lighters or matches. Do not cook without an adult present.*)
- What is an inferno? (*a large, out-of-control fire*)
- In an emergency, you can escape a building by opening any door marked with what kind of sign? (*EXIT*)

Activities, Supplies, and Preparations for this Lesson

- **Field trip:** Plan a field trip to the fire station and/or invite a fireman to visit your classroom to talk about fire safety.

- **Spelling Patterns of Sounds Heard in the English Language:** Display words with this week's sound on the wall display. (See pages 26-27 for details.)

- **Reproducible materials:** Prepare the reproducible materials provided with this lesson, including the question cards and the spelling word cards
(Supplies: Computer paper, cardstock)

- **Books and videos:** Read books and/or watch YouTube videos about firemen and fire safety. Also, plan to show children pictures of what firemen look like with their masks on so they know not to be afraid. These are available online. (For any books you read, identify the title, the author, and the illustrator. Also, examine the front and back cover of the book before making any predictions about the story or discussing what it is about.)

- **Fire safety coloring sheets:** Color the fire safety coloring sheets included with this lesson and read the captions.

- **EXIT signs:** Fold white construction paper in half and cut each piece into two rectangles from the crease. Give each child one rectangle, along with red cutout letters to spell the word *EXIT*. Show students an EXIT sign in the school building or online. Help them to understand that they could escape a building in the event of a fire or another emergency by going through any door marked with an Exit sign. Let students paste the letters to spell *Exit* onto their white paper.
(Supplies: Red and white construction paper; red cutout letters to spell *EXIT* for each child)
- **Stop, drop, and roll:** Let students practice the stop, drop, and roll technique with one of the songs available on YouTube. Explain that this technique smothers the oxygen that a fire needs to keep burning.

- **Crawl Under Smoke Activity:** Provide an inexpensive black or gray bedsheet for this activity. Assign a child to hold each corner of the sheet and position them near a door. Tell students that the sheet represents smoke and that smoke tends to rise. Then instruct them to crawl under the "smoke" and to feel the door for heat before opening it. Explain to students that it is difficult to see and breathe in a smoke-filled room. Tell them that smoke rises and that they can escape it by crawling under it.
(Supplies: An inexpensive black or gray bedsheet)

- **Fire Needs Oxygen Experiment:** Put a candle into a jar and light it. Then take away its oxygen supply by lying the lid across the opening. As the fire begins to go out, remove the lid to rekindle the fire. Explain to students that fire needs oxygen to keep burning. Put the lid back over the opening and let the fire go out.
(Supplies: Glass jar with a lid and a candle to fit inside; lighter to light the candle)

- **Writing/Handwriting assignment:** Position the handwriting assignment on page 148 under a Smart Document Camera or a similar device. Discuss the correct stroke for each letter as you copy the sentence onto your paper. Students should follow along with you, copying the letters on papers of their own. (This will help them to remain on task and to finish the handwriting assignment in unison.)

  Older Students: Older students can also write about the topic below. Let them copy the topic and respond to it in their writing journals. They may also want to add appropriate illustrations.

  Explain why you would or would not want to be a fireman.

Context Clues

Directions: Fill in each blank with a spelling word that makes sense in the context of the sentence. (Cross out the words from the list as you use them.) Circle the letters in the key words that make the /f/ sound (*f*, *fe*).

Spelling words: fire, force, fuel, face, fury, forest, fight, inferno, floor, flare, flames, safety

1. Fire is a ____________ of nature that can be helpful or destructive.
2. A ____________ fire can cause a lot of damage to trees.
3. Fortunately, we have firemen who can ____________ the fires.
4. When fires ____________ up, firemen are there to help us.
5. They are even brave enough to fight an ____________, which is a large, out-of-control fire.
6. The ____________ of an inferno can be deadly.
7. They work hard to put out the ____________.
8. Firemen want children to know about fire ____________.
9. Just like firewood, clothes can ____________ a fire.
10. If your clothes ever catch on ____________, do not run.
11. Drop to the ____________ immediately.
12. Then cover your ____________ and roll over and over again until the fire is out.

Spelling Words (/f/ sound)

F	A	C	E	B	I	N	F	E	R	N	O
U	K	L	I	L	K	S	U	T	F	B	X
F	U	O	D	M	T	O	F	L	O	O	R
O	T	F	U	R	Y	Y	V	D	X	H	Z
R	P	K	C	H	N	R	D	J	Z	D	A
E	I	F	F	J	F	L	A	R	E	N	F
S	A	O	Q	V	K	M	W	C	Z	S	L
T	F	R	Z	F	I	G	H	T	G	T	A
G	I	C	W	E	G	H	B	C	R	W	M
X	R	E	Y	W	M	F	U	E	L	Q	E
I	E	Q	E	I	P	X	Y	V	F	M	S
Q	Y	S	A	F	E	T	Y	J	P	J	X

Fire Safety Words

1. fire
2. force
3. fuel
4. face
5. fury
6. forest
7. fight
8. inferno
9. floor
10. flare
11. flames
12. safety

Compound Words

Directions: When two or more root words are used together to form one word, it is called a **compound word** (fire fighter). Complete each sentence by writing a word to form a compound in the blank space. The word must make sense in the context of the sentence.

1. A fire__________ is a person who fights fires.

   (hose, man, men, station, truck)

2. When fire__________ are notified of an emergency, they quickly leave the fire __________ and drive to the emergency in their fire__________.

   (hose, man, men, station, truck)

3. When they get to a fire, they use their fire__________ to put it out.

   (hose, man, men, station, truck)

Weekly Oral Language

1. Correct the asking sentence.

   what did you learn at school today

2. Correct the telling sentence.

   i learned about fire safety

3. What is the subject of the telling sentence?

   ____

4. Read the *telling sentence* again and identify the verb (action word).

   ____

5. Write your own sentence about this week's theme.

   ____

Instruction: *Teaching Children about Capitalization, Punctuation, Apostrophe s, and Contractions (Language Book 2)* by Habakkuk Educational Materials can be purchased separately to give students the knowledge they need to complete this worksheet. After students have completed the worksheet and turned it in, complete a copy of it on your classroom screen.

Parts of Speech

Directions: Identify what parts of speech are used in the sentence below and write them in the spaces provided. Then identify the subject (naming part) and predicate (action part) of the sentence and write them in the spaces provided.

Parts of speech: Noun marker, noun, pronoun, verb, helping verb, adjective, adverb, preposition, conjunction, interjection

I stopped, dropped, and rolled.

__________ __________ __________ __________ __________

Subject (Naming Part)

____

Predicate (Action Part)

____

Instruction: *Teaching Children about Parts of Speech (Language Book 3)* by Habakkuk Educational Materials can be purchased separately to help students quickly learn the parts of speech. After students have completed this worksheet and turned it in, complete a copy of it on your classroom screen.

Handwriting directions: Color the picture. Then copy the sentence below on the lines at the bottom of the page.

If your clothes are on fire, stop, drop, and roll.

Cover your mouth and nose
with clothing
to help protect you
against smoke inhalation.
paint the world
SUPER
COLORING

Fall & Crawl.
Stay low on the floor when escaping a fire.

Don't hide, go outside.

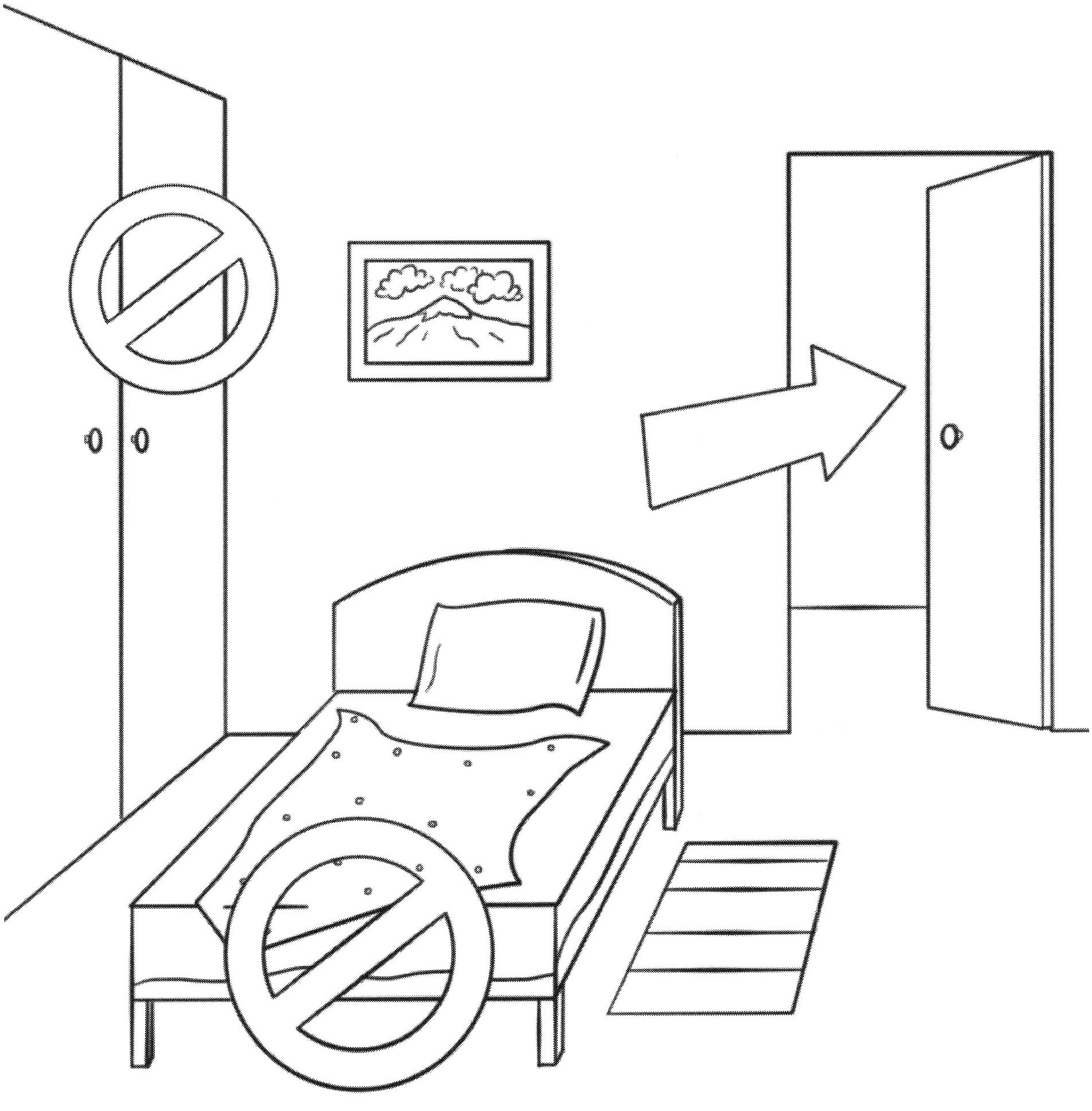

Call 911!

Test smoke alarms monthly to make sure they are working.

Replace old smoke alarm batteries with brand new ones at least once a year.

Try to find two ways out from every room in your home.

Never stick anything into an electrical socket.

Don't play with matches, lighters or candles.

Don't cook alone or without asking an adult.

Name: ____________________

Addition and Subtraction Facts

Directions

1. Count the firetrucks above the first space and write the addend.
2. Count the firetrucks above the second space and write the addend.
3. Count all the firetrucks and write the sum after the equal sign.

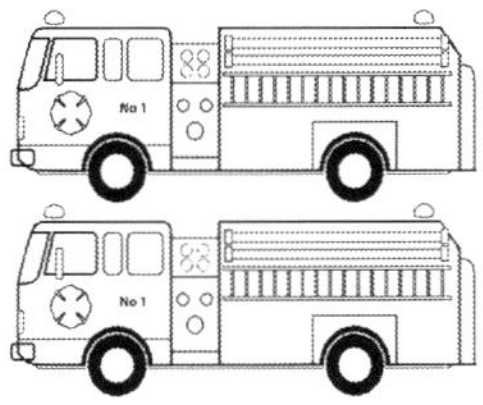

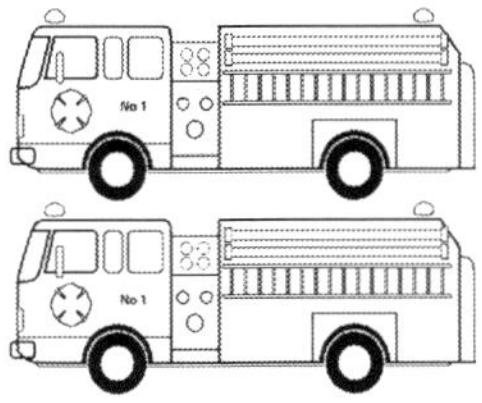

_______ + _______ = _______

Directions

1. Count the firetrucks above the first space and write the number.
2. Cross out 1 firetruck.
3. Count the number of firetrucks left and write the answer after the equal sign.

_______ − 1 = _______

QUESTIONS FOR THE THEMATIC UNITS GAME

Please refer to pages 28-29 of this book for instructions on how to use the unit questions.

What is the emergency number to dial if you ever need the police, an ambulance, or a fireman? *Lessons 7 (Child Safety)*	What does fire need to keep burning? *Lessons 7 (Child Safety)*
What would you do if your clothes caught on fire? *Lessons 7 (Child Safety)*	What would you do if you were making something to eat and burned your skin? *Lessons 7 (Child Safety)*
What would you do if the smoke alarm went off while you were home alone? *Lessons 7 (Child Safety)*	What precautions can you take to help protect you, your family, and your home from fires. *Lessons 7 (Child Safety)*
What are some things a child should never do because of the fire danger? *Lessons 7 (Child Safety)*	What is an inferno? *Lessons 7 (Child Safety)*
In an emergency, you can escape a building by opening any door marked with what kind of sign? *Lessons 7 (Child Safety)*	

UNIT QUESTIONS FOR NEXT WEEK'S LESSON

Where was Christopher Columbus born? *Lesson 8 (Social Studies)*	When Christopher Columbus was a young man, he left his home and moved to Portugal. Later, he left Portugal and moved where? *Lesson 8 (Social Studies)*
Who provided the money and ships so that Christopher Columbus could make his famous voyage? *Lesson 8 (Social Studies)*	How many men sailed with Christopher Columbus on his first voyage to America? *Lesson 8 (Social Studies)*
What were the names of the three ships that Christopher Columbus and the other sailors sailed on to America? *Lesson 8(Social Studies)*	Christopher Columbus was not looking for America when he happened upon it. What was he looking for? *Lesson 8 (Social Studies)*
On October 12, 1492, Christopher Columbus landed on an island near to present-day Florida and found many Natives living there. Why did he call them Indians? *Lesson 8 (Social Studies)*	Why do many Americans celebrate Columbus Day the second Monday of October every year? *Lesson 8 (Social Studies)*

LESSON 7 TEST (Fire Safety)

1. What is the emergency number to dial if you ever need the police, an ambulance, or a fireman?

   ○ 000

   ○ 911

   ○ 999

2. What does fire need to keep burning?

   ○ Oxygen (or air)

   ○ Water

   ○ Smoke

3. What should you do if your clothes catch on fire?

   ○ Run as fast as you can.

   ○ Stop, drop, and roll.

   ○ Jump up and down.

4. What would you do if you were making something to eat and burned your skin?

   ○ Put the burn in ice water and call the emergency number if the burn is serious.

   ○ Put the burn in cool water (not ice water) and call the emergency number if the burn is serious.

   ○ Fan the burn to cool it off.

5. What would you do if the smoke alarm went off while you were home alone?

   ○ First get out of the house. Then call the emergency number.

   ○ First call the emergency number. Then get out of the house.

   ○ Hide in the closet.

6. What is the safest way to escape your house if you hear the smoke alarm sounding? (Mark all that apply.)

- ○ Feel the bedroom door to make sure it is not hot before opening it. If it is hot or if the doorknob feels hot, crawl out the window instead.
- ○ Crawl under any smoke to get to a door or window.
- ○ Use your shirt or something else to cover your nose and your mouth so that you don't inhale the smoke.

7. What precautions can you take to help protect you, your family, and your home from fires?

- ○ Check the smoke alarms monthly and replace the batteries yearly.
- ○ Practice crawling out of your bedroom window.
- ○ Both of the above answers are true.

8. What are some things a child should never do because of the fire danger? (Mark all that apply.)

- ○ Never stick anything into a light socket.
- ○ Do not play with lighters or matches.
- ○ Do not cook without an adult present.

9. What is an inferno?

- ○ A lit match
- ○ A campfire
- ○ A large, out-of-control fire

10. In an emergency, you can escape a building by opening any door marked with which sign?

- ○ CLOSED
- ○ EXIT
- ○ OPEN

Note: These cards can be used for the spelling word games listed on pages 21-22.

fire	force
fuel	face
fury	forest

Note: These cards can be used for the spelling word games listed on pages 21-22.

fight	inferno
floor	flare
flames	safety

Lesson 8

Theme: Columbus Day
Sound: short *i* /ĭ/
Spelling Patterns: *i*, *e*, *a*

Spelling Words	Alphabetized
1. Italy	________________
2. Indies	________________
3. Indians	________________
4. ships	________________
5. trip	________________
6. give	________________
7. gifts	________________
8. Natives	________________
9. discovery	________________
10. embark	________________
11. voyage	________________
12. courage	________________

Inflectional endings: embarks, embarking, embarked

Extra challenge word: Christopher

Note: The initial letters of *Christopher*, *Italy*, *Indies*, and *Indian* must be capitalized.

Thematic Activities for Lesson 8
(Columbus Day)

Thematic objectives: Students will learn some facts about Christopher Columbus, his journey to America, and why we celebrate Columbus Day.

Key Questions for Discussion

- Where was Christopher Columbus born? (*Italy*)
- When Christopher Columbus was a young man, he left his home and moved to Portugal. Later, he left Portugal and moved where? (*to Spain*)
- Who provided the money and ships so that Christopher Columbus could make his famous voyage? (*the king and queen of Spain*)
- How many men sailed with Christopher Columbus on his first voyage to America? (*almost 90*)
- What were the names of the three ships that Christopher Columbus and the other sailors sailed on to America? (*Niña, Pinta, and Santa Maria*)
- Christopher Columbus was not looking for America when he happened upon it. What was he looking for? (*He wanted to find a quicker route to the Indies to look for jewels and other valuables.*)
- On October 12, 1492, Christopher Columbus landed on an island near to present-day Florida and found many Natives living there. Why did he call them Indians? (*He thought he had found the Indies.*)
- Why do many Americans celebrate Columbus Day the second Monday of October every year? (*to remember the discovery of America*)

Activities, Supplies, and Preparations for this Lesson

- **Spelling Patterns of Sounds Heard in the English Language:** Display words with this week's sound on the wall display. (See pages 26-27 for details.)
- **Reproducible materials:** Prepare the reproducible materials provided with this lesson, including the question cards and the spelling word cards (Supplies: Computer paper, cardstock)
- **Books and videos:** Read books and/or watch YouTube videos about Christopher Columbus. (For any books you read, identify the title, the author, and the illustrator. Also, examine the front and back cover of the book before making any predictions about the story or discussing what it is about.) (Suggested YouTube Video: *A Picture Book of Christopher Columbus*, by David A. Adler, video by Bookworm Bonny)

- **Geography:** Use a world map to show students where Christopher Columbus was born (Italy—the country that looks like a boot), where he moved to as a young man and eventually set sail from on his voyage to America (Spain), and where he eventually landed (the Bahamas near to present-day Florida).
- **Thematic Units Game for Christopher Columbus:** Provide three pictures of ships to represent the *Niña*, *Pinta*, and *Santa Maria* that Christopher Columbus and the other sailors traveled on to America. Tape each image to a lunch bag (one picture per bag) and record the name of the ship below the picture. You may want to use wall-safe tape to attach the bags to the wall. Reproduce the unit questions from page 160 on cardstock, cut apart the pieces, and distribute them between the bags. Question cards from previous lessons can also be included in the bags. As a variation to using the unit questions, the spelling word cards can also be used with the game. Record the names of the ships in the spaces of a blank spinner. (Blank spinners are available on the "Free Teaching Materials" page of the Habakkuk Educational Materials website. Click on "Board Games and More" to access the spinners.) A student spins the spinner, reads the name of the ship that the spinner lands on, and then draws a card from the lunch bag with the picture representing the ship. If he answers the question or reads the spelling word correctly, his team receives one point. (Supplies: Cardstock; a brad and a large paperclip; lunch bags; three pictures of ships to represent the *Niña*, *Pinta*, and *Santa Maria*; wall-safe tape)
- **Writing/Handwriting assignment:** Position the handwriting assignment on page 174 under a Smart Document Camera or a similar device. Discuss the correct stroke for each letter as you copy the sentence onto your paper. Students should follow along with you, copying the letters on papers of their own. (This will help them to remain on task and to finish the handwriting assignment in unison.)

  Older Students: Older students can also write about the topic below. Let them copy the question and respond to it in their writing journals. They may also want to add appropriate illustrations.

  What have you learned about Christopher Columbus?

Context Clues

Directions: Fill in each blank with a spelling word that makes sense in the context of the sentence. (Cross out the words from the list as you use them.) Circle the letter in the key words that make the short *i* sound (*i, e, a*).

Spelling words: Italy, Indies, Indians, ships, trip, give, gifts, Natives, discovery, embark, voyage, courage

1. Christopher Columbus was born in ___________ but moved to Spain when he was a young man.

2. He was a sailor and wanted to find a quicker route to the ___________ to look for jewels and other valuables.

3. He asked the king and queen of Spain for money and ships to make the ___________, but they said no.

4. Years later, however, in 1492, the king and queen of Spain changed their minds and decided to ___________ Columbus three ships.

5. The name of the ___________ were the Niña, Pinta, and Santa Maria.

6. Almost 90 men went on the ___________ with Christopher Columbus.

7. It must have been a scary trip, but Columbus had a lot of __________.

8. On October 12, 1492, they landed on an island near to present-day Florida and found many __________ living there.

9. Christopher Columbus called the people __________ because he thought he had found the Indies.

10. Columbus returned to Spain with __________ from the New World.

11. The king and queen were glad they allowed Columbus to __________ on his journey.

12. Every year on the second Monday of October, Americans celebrate Columbus Day to remember the __________ of America.

Spelling Patterns of Sounds Heard in Our Spelling Words

Directions: Fill in the missing letters that make the short *i* sound /ĭ/.

- Use a **blue** colored pencil for ***i***.
- Use a **red** colored pencil for ***e***.
- Use a **green** colored pencil for ***a***.

_taly	tr_p	d_scovery
_ndies	g_ve	_mbark
_ndians	g_fts	voy_ge
sh_ps	Nat_ves	cour_ge

Directions: Record spelling words whose short *i* sound is spelled *i* under the **"i"** headings, any word whose /ĭ/ sound is spelled *e* under the **"e"** heading, and words whose /ĭ/ sound is spelled *a* under the **"a"** heading.

i	i	e	a
________	________	________	________
________	________	________	________
________	________	________	________
________	________	________	________
________	________	________	________

Spelling Words (short *i* /ĭ/ sound)

N	C	N	V	I	N	D	I	A	N	S	O
A	S	V	I	L	K	S	U	T	F	B	D
T	U	O	D	I	T	A	L	Y	T	J	I
I	T	Y	M	H	T	Y	V	D	X	H	S
V	P	A	C	G	I	V	E	J	E	D	C
E	I	G	F	J	D	O	S	L	M	N	O
S	A	E	Q	V	K	M	H	C	B	S	V
A	G	I	F	T	S	H	I	V	A	T	E
G	S	L	W	E	G	H	P	C	R	W	R
I	N	D	I	E	S	E	S	H	K	Q	Y
I	R	Q	E	I	P	X	Y	K	F	M	R
T	R	I	P	B	C	O	U	R	A	G	E

Columbus Day

1. Italy
2. Indies
3. Indians
4. ships
5. trip
6. give
7. gifts
8. Natives
9. discovery
10. embark
11. voyage
12. courage

Weekly Oral Language

1. Correct the asking sentence.

   what were the names of the three ships that columbus and the other sailors traveled on to america

2. Correct the telling sentence.

   the ships were named the Niña, Pinta, and Santa Maria

3. What is the subject of the telling sentence?

   ____

4. Read the *telling sentence* again and identify the helping verb and the verb (action word).

   Helping Verb: ____________________________

   Verb: ____________________________________

5. Write your own sentence about this week's theme.

   ____

Instruction: *Teaching Children about Capitalization, Punctuation, Apostrophe s, and Contractions (Language Book 2)* by Habakkuk Educational Materials can be purchased separately to give students the knowledge they need to complete this worksheet. After students have completed the worksheet and turned it in, complete a copy of it on your classroom screen.

Parts of Speech

Directions: Identify what parts of speech are used in the sentence below and write them in the spaces provided. Then identify the subject (naming part) and predicate (action part) of the sentence and write them in the spaces provided.

Parts of speech: Noun marker, noun, pronoun, verb, helping verb, adjective, adverb, preposition, conjunction, interjection

Columbus	**Day**	**is**	**in**	**October.**
________	________	________	________	________

Subject (Naming Part)

____

Predicate (Action Part)

____

Instruction: *Teaching Children about Parts of Speech (Language Book 3)* by Habakkuk Educational Materials can be purchased separately to help students quickly learn the parts of speech. After students have completed this worksheet and turned it in, complete a copy of it on your classroom screen.

Handwriting directions: Color the picture. Then copy the sentence below on the lines at the bottom of the page.

On October 12, 1492, Christopher Columbus landed on an Island near to present-day Florida in America.

LESSON 8 TEST (Columbus Day)

1. Where was Christopher Columbus born?

- ◯ America
- ◯ Italy
- ◯ Spain

2. When Christopher Columbus was a young man, he left his home and moved to Portugal. Later, he left Portugal and moved where?

- ◯ America
- ◯ Italy
- ◯ Spain

3. Who provided the money and ships so that Christopher Columbus could make his famous voyage?

- ◯ Christopher Columbus' brother
- ◯ The king and queen of Spain
- ◯ The king of England

4. How many men sailed with Christopher Columbus on his first voyage to America?

- ◯ Almost 90
- ◯ Almost 50
- ◯ Almost 200

5. What were the names of the three ships that Christopher Columbus and the other sailors traveled on to America?

- ◯ Discovery, Susan Constant, and Godspeed
- ◯ The Mayflower, the Titanic, and the Mary Rose
- ◯ Niña, Pinta, and Santa Maria

6. What was Christopher Columbus looking for when he and the other sailors set sail in the 3 ships?

- ◯ A quicker route to America to look for jewels and other valuables
- ◯ A quicker route to the Indies to look for jewels and other valuables
- ◯ A quicker route to Italy to look for jewels and other valuables

7. On October 12, 1492, Christopher Columbus landed on an island near to present-day Florida and found many Natives living there. Why did he call them Indians?

- ◯ He thought he had found the Indies.
- ◯ He knew he had found America.
- ◯ He thought they would like that name.

8. Why do many Americans celebrate Columbus Day the second Monday of October every year?

- ◯ To celebrate Columbus' birthday
- ◯ To remember the discovery of the Indies
- ◯ To remember the discovery of America

Note: These cards can be used for the spelling word games listed on pages 21-22.

Italy	Indies
Indians	ships
trip	give

Note: These cards can be used for the spelling word games listed on pages 21-22.

gifts	Natives
discovery	embark
voyage	courage

Lesson 9

Theme: The Fall Season (Autumn)
Sound: Short *o* /ŏ/
Spelling Patterns: *o, a, au, aw*

Spelling Words	Alphabetized
1. drop	______________
2. off	______________
3. logs	______________
4. flock	______________
5. crops	______________
6. fall	______________
7. football	______________
8. stalk	______________
9. pecan	______________
10. autumn	______________
11. lawn	______________
12. saw	______________

Inflectional endings: saws, sawing, sawed

Bonus compound words: cornstalk, lawnmower

Extra challenge word: October (The initial letter must be capitalized.)

Thematic Activities for Lesson 9
(The Fall Season)

Thematic objectives: Students will understand that fall is one of the four seasons and will be able to identify animal behavior, weather conditions, and activities that occur in autumn.

Key Questions for Discussion

- Fall is one of the four seasons. What are the four seasons? (*winter, spring, summer, fall*)
- What happens in nature during the fall season? (*The weather turns cooler, and leaves turn pretty colors and fall off the trees.*)
- What do farmers usually do in the fall? (*harvest crops they planted in the spring*)
- What do squirrels do in the fall? (*They hide acorns and other nuts that drop from the trees in fall so they have something to eat during winter.*)
- What do many birds do in the fall? (*Many birds fly south in fall to get to warmer weather.*)
- What are the fall months? (*Part of September, all of October and November, part of December*)
- What is another name for the season fall? (*autumn*)

Activities, Supplies, and Preparations for this Lesson

- **Spelling Patterns of Sounds Heard in the English Language:** Display words with this week's sound on the wall display. (See pages 26-27 for details.)

- **Reproducible materials:** Prepare the reproducible materials provided with this lesson, including the "Spelling Word Riddles," the question cards, and the spelling word cards
  (Supplies: Computer paper, light blue and yellow cardstock, magnetic tape)

- **Books and videos:** Read books and/or watch YouTube videos about seasons and autumn. Have Fun Teaching has a cute video available on YouTube titled "Seasons Song" that can help students learn about the four seasons. (For any books you read, identify the title, the author, and the illustrator. Also, examine the front and back cover of the book before making any predictions about the story or discussing what it is about.)

- **Leaf Rubbing:** Give your students a lunch bag and take them on a nature walk to collect leaves. Instruct students to put a piece of paper atop a leaf and to rub over it with a yellow, red, orange, or brown crayon to create fall colored leaves.
  (<u>Supplies:</u> Lunch bags; plain, white paper; yellow, red, orange, and brown crayons)

- **Writing/Handwriting assignment:** Position the handwriting assignment on page 193 (i.e., the page with the squirrel) under a Smart Document Camera or a similar device. Discuss the correct stroke for each letter as you copy the sentence onto your paper. Students should follow along with you, copying the letters on papers of their own. (This will help them to remain on task and to finish the handwriting assignment in unison.)

  <u>Older Students:</u> Older students can also write about the topic below. Let them copy the question and respond to it in their writing journals. They may also want to add appropriate illustrations.

  What is your favorite thing about the fall season and why?

Context Clues

Directions: Fill in each blank with a spelling word that makes sense in the context of the sentence. (Cross out the words from the list as you use them.) Circle the letters in the key words that make the short *o* sound (*o*, *a*, *au*, *aw*).

Spelling words: drop, off, logs, flock, crops, fall, football, stalk, pecan, autumn, lawn, saw

1. ____________ is one of the four seasons.

2. Another word for fall is ____________.

3. Farmers harvest their ____________ in the fall.

4. They harvest corn by removing it from its ____________.

5. The leaves turn pretty colors and fall ____________ the trees.

6. It's fun to jump in all the leaves after raking the ____________.

7. Acorns also ____________ from the trees, and squirrels hide them so they have something to eat during winter.

8. The ____________ is another nut that falls from trees in autumn.

9. The weather starts turning cooler in autumn, so you may see a ____________ of geese flying south where it is warm.

10. We put ____________ in our fireplace to keep warm.

11. Sometimes, Dad lets me help ____________ the logs.

12. Dad loves the fall because he gets to watch a lot of ____________ games.

Spelling Word Riddles

Discussion: All spelling words this week have the short *o* sound, /ŏ/, and can be spelled as ***o*** as in *drop*, ***a*** as in *fall*, ***au*** as in *autumn*, or ***aw*** as in *lawn*. In the word *autumn*, what vowel sound do you hear? (Help students hear and pronounce the short *o* sound by first saying the sound and then a basic word, such as *autumn*.)

Directions: Reproduce the next two pages on light blue or yellow cardstock as instructed. Cut out the pieces and attach a magnetic strip to the back of each piece. Line up the yellow pieces horizontally towards the top of a magnetic board. Position the light blue strips vertically below them. (See the illustration on page 15.) Read a riddle and call on a student to identify the correct spelling word. For the first sentence, a student would identify the word *autumn* and then move the yellow magnetic strip with the letters ***au*** to the left of the light blue magnetic strip having the letters ***tumn*** to form the word ***autumn***.

1. In this word, the short *o* sound is spelled *a-u*. This season comes after summer. (*Autumn*)
2. In this word, the short *o* sound is spelled *a*. This is a sport a lot of boys play. (*Football*)
3. In this word, the short *o* sound is spelled *a-w*. This needs to be raked a lot in the fall. (*Lawn*)
4. In this word, the short *o* sound is spelled *o*. We burn these in the fireplace to keep warm. (*Logs*)
5. In this word, the short *o* sound is spelled *o*. Farmers harvest these in the fall. (*Crops*)
6. In this word, the short *o* sound is spelled *o*. It can refer to a group of birds flying south together. (*Flock*)
7. In this word, the short *o* sound is spelled *a*. Corncobs grow on this. (*Stalk*)
8. In this word, the short *o* sound is spelled *a*. You might see a squirrel eating this nut. (*Pecan*)
9. In this word, the short *o* sound is spelled *a*. This is another word for Autumn. (*Fall*)
10. In this word, the short *o* sound is spelled *a-w*. It can be used to cut firewood. (*Saw*)
11. In this word, the short *o* sound is spelled *o*. It means the opposite of *on*. (*Off*)
12. In this word, the short *o* sound is spelled *o*. It rhymes with *crop*. (*Drop*)

Directions: Use light blue cardstock to reproduce this page. Cut out the pieces and attach magnetic strips to the back of each piece. (See the previous page for further instructions.)

dr	p	ff	
l	gs	fl	ck
cr	ps	f	ll
footb	ll	st	lk
pec	n	tumn	
l	n	s	

Directions: Use yellow cardstock to reproduce this page. Cut out the pieces and attach magnetic strips to the backs.

au	au	au	au	au
aw	aw	aw	aw	aw

o	o	o	o	o	a	a	a	a	a

Spelling Patterns of Sounds Heard in Our Spelling Words

Directions: Fill in the missing letters that make the short *o* sound /ŏ/.

- Use a **blue** colored pencil for ***o***.
- Use a **red** colored pencil for ***a***.
- Use a **green** colored pencil for ***au***.
- Use a **purple** colored pencil for ***aw***.

d r _ p	c r _ p s	p e c _ n
_ f f	f _ l l	_ _ t u m n
l _ g s	f o o t b _ l l	l _ _ n
f l _ c k	s t _ l k	s _ _

Directions: Record spelling words whose short *o* sound is spelled *o* under the **"o"** heading, words whose /ŏ/ sound is spelled *a* under the **"a"** heading, and so forth.

o	a	au	aw
____________	____________	____________	____________
____________	____________	____________	____________
____________	____________	____________	____________
____________	____________	____________	____________
____________	____________	____________	____________

Spelling Words (Short *o* /ŏ/ sound)

G	S	T	A	L	K	A	O	F	F	U	F
U	K	L	I	L	K	S	U	T	F	B	O
D	U	O	D	M	T	L	A	W	N	J	O
R	T	F	A	L	L	Y	V	D	X	H	T
O	P	K	C	H	N	R	D	J	L	D	B
P	I	N	F	J	D	O	P	L	O	N	A
S	D	A	Q	V	K	M	E	C	G	S	L
X	F	N	Z	Q	F	H	C	K	S	T	L
A	C	R	O	P	S	H	A	C	R	W	O
G	G	Y	Y	W	M	E	N	H	S	A	W
F	L	O	C	K	P	X	Y	V	F	M	R
Q	Y	L	F	B	Z	A	U	T	U	M	N

Autumn Words

1. drop
2. off
3. logs
4. flock
5. crops
6. fall
7. football
8. stalk
9. pecan
10. autumn
11. lawn
12. saw

Homophones

stalk

Corn grows on a **stalk**.

A stalk is part of a plant and is similar to a stem.

stock

Squirrels **stock** up nuts for the winter.

Stock has to do with gathering a lot of something and storing it away.

In fall, squirrels **stock** up nuts, and farmers remove corn from the **stalk**.

Underline the word that refers to part of a plant.
Circle the word that has to do with gathering a lot of something and storing it away.

Compound Words

Directions: When two or more root words are used together to form one word, it is called a **compound word** (corncob). Form three compound words by adding a word from the word list to the words *corn*, *foot*, and *mower*.

Word list: ball, lawn, stalk

corn________ foot________ ________mower

Directions: Fill in each blank with one of the compound words that makes sense in the context of the sentence.

Today, I'm going to help my grandfather gather corn from the ________________. Then we're going to use the ________________ to mow over all the leaves in the yard. After that, we get to watch the ________________ game!

Weekly Oral Language

1. Correct the asking sentence.

   where do pecans grow

2. Correct the telling sentence.

   pecans grow on trees

3. What is the subject of the telling sentence?

   ____

4. Read the *telling sentence* again and identify the verb (action word).

   ____

5. Write your own sentence about this week's theme.

   ____

Instruction: *Teaching Children about Capitalization, Punctuation, Apostrophe s, and Contractions (Language Book 2)* by Habakkuk Educational Materials can be purchased separately to give students the knowledge they need to complete this worksheet. After students have completed the worksheet and turned it in, complete a copy of it on your classroom screen.

Parts of Speech

Directions: Identify what parts of speech are used in the sentence below and write them in the spaces provided. Then identify the subject (naming part) and predicate (action part) of the sentence and write them in the spaces provided.

Parts of speech: Noun marker, noun, pronoun, verb, helping verb, adjective, adverb, preposition, conjunction, interjection

Bears,	**bats,**	**and**	**skunks**	**hibernate.**
________	________	________	________	________

Subject (Naming Part)

____

Predicate (Action Part)

____

Instruction: *Teaching Children about Parts of Speech (Language Book 3)* by Habakkuk Educational Materials can be purchased separately to help students quickly learn the parts of speech. After students have completed this worksheet and turned it in, complete a copy of it on your classroom screen.

Short *O* Sound /Ŏ/

Directions: Write letters in the spaces provided to form words with the key sound. Call on students to sound out the words. Then use one or more of the words in a sentence.

Audio	S**aw**
____au____	____aw____
____au____	____aw____
F**al**l, W**al**k	**Wa**sh
____al____	____a____
____al____	____a____

____

Rule: The vowel *o* often makes the short *u* sound (a schwa sound), /ə/, when it's followed by *m*, *n*, or *v* and is referred to as a "scribal *o*."

Examples: c**o**m**e**, fr**o**m, d**o**n**e**, t**o**n, ab**o**v**e**, l**o**v**e**

Rule: The letter combination *all* makes the /ȯl/ sound when it ends a word. (Fall)

Examples: ***au*** (August, haul, launch, sauce); ***aw*** (draw, jaw, law, raw); ***al*** (all, ball, call, small, balk, chalk, stalk, talk); ***a*** (wad, wasp, watch, water)

Handwriting directions: Color the squirrel. Then copy the sentence below on the lines at the bottom of the page.

In autumn, squirrels hide acorns and other nuts to eat during winter.

Name: ______________________________

Addition and Subtraction Facts

Directions

1. Count the leaves above the first space and write the addend.
2. Count the leaves above the second space and write the addend.
3. Count all the leaves and write the sum after the equal sign.

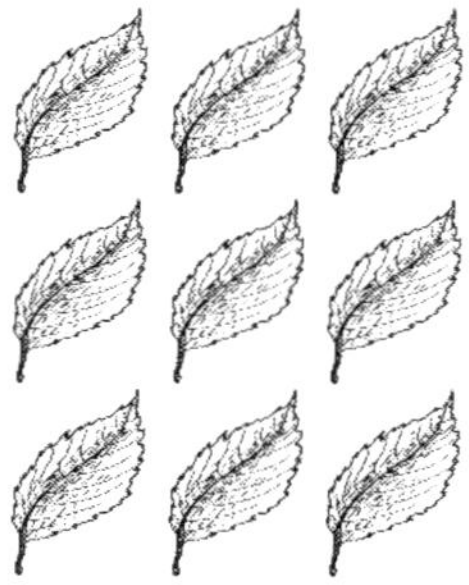

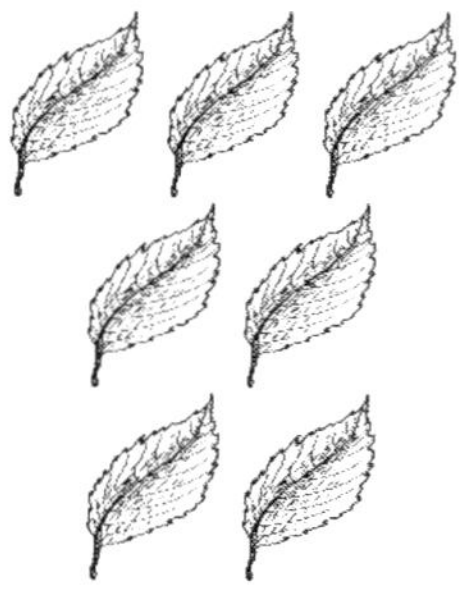

______ + ______ = ______

Directions

1. Count the acorns above the first space and write the number.
2. Cross out 7 acorns.
3. Count the number of acorns left and write the answer after the equal sign.

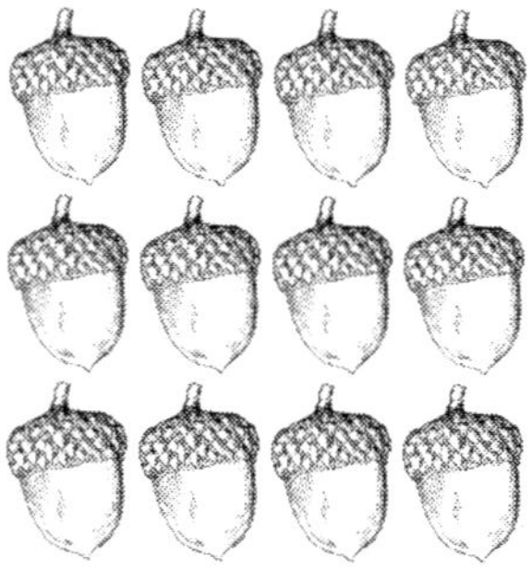

______ − __7__ = ______

Subtraction Word Problems

Directions: Color and cut out the pictures at the bottom of the page. Use autumn colors (red, orange, brown, and yellow) for the leaves. Read each sentence and use the cutouts to help solve the problem. Write the subtraction fact in the spaces provided.

1. There were 8 leaves on the tree. Five leaves fell off. How many leaves are still on the tree?

   _____ – _____ = _____

2. There were 5 apples on the tree. Beau picked 1. How many apples are still on the tree?

   _____ – _____ = _____

3. A group of birds were flying South for the winter when 9 of them stopped to rest on my tree. If 2 birds already took off, how many birds are still on my tree?

   _____ – _____ = _____

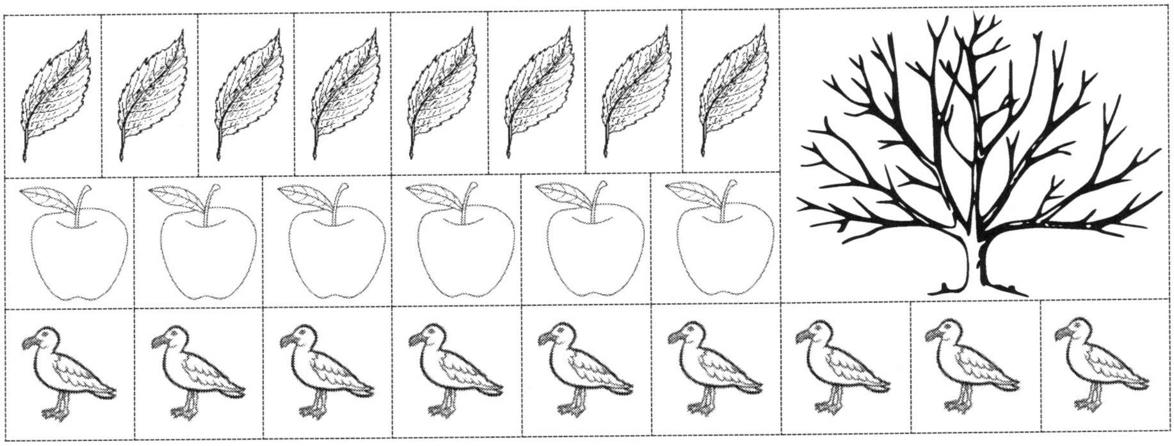

QUESTIONS FOR THE THEMATIC UNITS GAME

Please refer to pages 28-29 of this book for instructions on how to use the unit questions.

Fall is one of the four seasons. What are the four seasons? *Lesson 9 (Science Themes)*	What happens in nature during the fall season? *Lesson 9 (Science Themes)*
What do farmers usually do in the fall? *Lesson 9 (Science Themes)*	What do squirrels do in the fall? *Lesson 9 (Science Themes)*
What do many birds do in the fall? *Lesson 9 (Science Themes)*	What are the fall months? *Lesson 9 (Science Themes)*
What is another name for the season fall? *Lesson 9 (Science Themes)*	

LESSON 9 TEST (The Fall Season)

1. What are the four seasons?
   - ◯ Sunday, Monday, Tuesday, and Wednesday
   - ◯ September, October, November, and December
   - ◯ Winter, spring, summer, and fall

2. What happens in nature during the fall season?
   - ◯ The weather turns cooler.
   - ◯ Leaves turn pretty colors and fall off the trees.
   - ◯ Both of the above answers are true.

3. What do farmers usually do in the fall?
   - ◯ Plant their crops
   - ◯ Harvest their crops
   - ◯ Take a long vacation

4. What do animals do in fall?
   - ◯ Many birds fly south where it is warmer.
   - ◯ Squirrels hide acorns and other nuts to eat during winter.
   - ◯ Both of the above answers are true.

5. What are the fall months?
   - ◯ March, April, and May
   - ◯ June, July, and August
   - ◯ September, October, and November

6. What is another name for the season fall?

- ◯ Winter
- ◯ Autumn
- ◯ October

Note: These cards can be used for the spelling word games listed on pages 21-22.

dr<u>o</u>p	<u>o</u>ff
l<u>o</u>gs	fl<u>o</u>ck
cr<u>o</u>ps	f<u>a</u>ll

Note: These cards can be used for the spelling word games listed on pages 21-22.

football	stalk
pecan	autumn
lawn	saw

Lesson 10

Theme: It's Fun to Pretend (Halloween)
Sound: short *a* /ă/
Spelling Pattern: *a*

Spelling Words	Alphabetized
1. hat	____________
2. shadow	____________
3. apple	____________
4. bats	____________
5. bag	____________
6. mask	____________
7. black	____________
8. cat	____________
9. Jack	____________
10. lantern	____________
11. candy	____________
12. candle	____________

Bonus compound words: flashlight, jack-o'-lantern

Extra challenge words: Halloween (The initial letter must be capitalized.)

Context Clues

Directions: Fill in each blank with a spelling word that makes sense in the context of the sentence. (Cross out the words from the list as you use them.) Circle the letter in the key words that make the short *a* sound /ă/.

Spelling words: hat, shadow, apple, bats, bag, mask, black, cat, Jack, lantern, candy, candle

1. For Halloween, I dressed up in western clothes and wore a cowboy ____________.
2. My sister was dressed like a superhero and got to wear an eye ____________.
3. Our cousin dressed all in black to look like someone's ____________.
4. We went to a party and saw ____________ hanging from the ceiling.
5. They also had a stuffed black ____________ but nothing scary.
6. There were a lot of orange and ____________ balloons.
7. There was also a ____________-o'-Lantern contest.
8. Every pumpkin had a ____________ inside so you could see it glowing.
9. The best jack-o'-____________ won a prize.
10. I ate a caramel ____________ and played a lot of games.
11. We got ____________ for playing the games.
12. I got a lot of candy in my ____________!

Spelling Words (Short *a* /ă/ sound)

K	C	N	C	A	N	D	Y	E	A	U	L
U	A	L	I	L	K	S	U	T	P	B	A
G	T	O	D	M	R	A	O	E	P	J	N
S	T	B	L	A	C	K	V	D	L	H	T
H	P	K	C	H	N	R	D	J	E	D	E
A	I	B	F	J	D	O	A	L	M	N	R
D	A	A	Q	J	A	C	K	C	Z	S	N
O	F	T	Z	Q	F	H	P	K	M	T	E
W	S	S	W	E	G	H	B	C	A	W	O
X	G	Y	Y	W	M	E	P	H	S	Q	H
C	A	N	D	L	E	X	Y	V	K	M	A
Q	Y	L	F	B	Z	B	A	G	P	J	T

Halloween Words

1. hat
2. shadow
3. apple
4. bats
5. bag
6. mask
7. black
8. cat
9. Jack
10. lantern
11. candy
12. candle

Weekly Oral Language

1. Correct the asking sentence.

   what are you eating

2. Correct the telling sentence.

   i am eating my halloween candy

3. What is the subject of the telling sentence?

   ____

4. Read the *telling sentence* again and identify the helping verb and the verb (action word).

   Helping Verb: ______________________

   Verb: ______________________________

5. Write your own sentence about this week's theme.

   ____

Instruction: *Teaching Children about Capitalization, Punctuation, Apostrophe s, and Contractions (Language Book 2)* by Habakkuk Educational Materials can be purchased separately to give students the knowledge they need to complete this worksheet. After students have completed the worksheet and turned it in, complete a copy of it on your classroom screen.

Parts of Speech

Directions: Identify what parts of speech are used in the sentence below and write them in the spaces provided. Then identify the subject (naming part) and predicate (action part) of the sentence and write them in the spaces provided.

Parts of speech: Noun marker, noun, pronoun, verb, helping verb, adjective, adverb, preposition, conjunction, interjection

We	**are**	**bobbing**	**for**	**apples.**
__________	__________	__________	__________	__________

Subject (Naming Part)

____

Predicate (Action Part)

____

Instruction: *Teaching Children about Parts of Speech (Language Book 3)* by Habakkuk Educational Materials can be purchased separately to help students quickly learn the parts of speech. After students have completed this worksheet and turned it in, complete a copy of it on your classroom screen.

Name: ____________________

Addition and Subtraction Facts

Directions

1. Count the bats above the first space and write the addend.
2. Count the bats above the second space and write the addend.
3. Count all the bats and write the sum after the equal sign.

________ + ________ = ________

Directions

1. Count the pumpkins above the first space and write the number.
2. Cross out 6 pumpkins.
3. Count the number of pumpkins left and write the answer after the equal sign.

________ − 6 = ________

Addition and Subtraction Word Problems

Teacher instructions for a center activity: Provide wrapped candy and a Halloween bucket. Students color and cut out the pumpkins at the bottom of the page. Instruct them to read the sentences, to record the equations, and to use the candy and cutouts to help solve the problems.

1. There were 14 pumpkins in the pumpkin patch. A family took 8 home. How many pumpkins are left in the patch?

   ___________ = _____

2. I had 10 pieces of candy in my Halloween bucket. I ate 2 pieces. How many pieces are still in my bucket?

   ___________ = _____

3. Our class has 13 pumpkins to decorate. Six pumpkins have already been decorated. How many pumpkins do we still need to decorate?

   ___________ = _____

4. I had 7 pieces of candy in my Halloween bag. My neighbor put in 2 more pieces. How many pieces are in my bag now?

   ___________ = _____

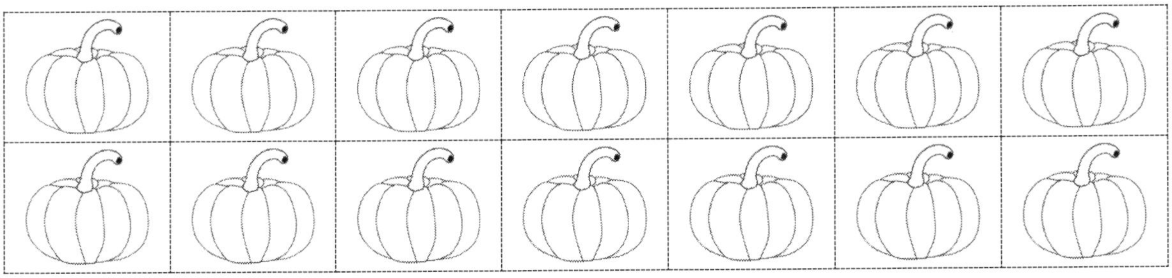

Note: These cards can be used for the spelling word games listed on pages 21-22.

hat	shadow
apple	bats
bag	mask

Note: These cards can be used for the spelling word games listed on pages 21-22.

bl<u>a</u>ck	c<u>a</u>t
J<u>a</u>ck	l<u>a</u>ntern
c<u>a</u>ndy	c<u>a</u>ndle

Lesson 11

Theme: Pets (Domesticated Animals)
Sound: /ĕ/
Spelling Patterns: *e*, *ea*

Spelling Words	Alphabetized
1. pet	________________
2. bed	________________
3. beg	________________
4. fed	________________
5. fetch	________________
6. vet	________________
7. smell	________________
8. help	________________
9. rest	________________
10. sell	________________
11. met	________________
12. head	________________

Inflectional endings: fetches, fetching, fetched

Thematic Activities for Lesson 11
(Pets)

Thematic objectives: Students will learn that pets like dogs and cats are examples of domesticated animals because they rely upon humans for survival.

Key Questions for Discussion

- What is a domesticated animal? (*an animal that relies upon humans for survival*)
- What are examples of domesticated animals? (*a pet like a dog or cat and farm animals that rely upon humans for food and other things they need)*

Activities, Supplies, and Preparations for this Lesson

- **Spelling Patterns of Sounds Heard in the English Language:** Display words with this week's sound on the wall display. (See pages 26-27 for details.)

- **Reproducible materials:** Prepare the reproducible materials provided with this lesson, including the question cards and the spelling word cards (<u>Supplies:</u> Computer paper, cardstock)

- **Books and videos:** Read books and/or watch YouTube videos about pets and domesticated animals, especially dogs and cats. (For any books you read, identify the title, the author, and the illustrator. Also, examine the front and back cover of the book before making any predictions about the story or discussing what it is about.)

- **Discussion:** Give students the opportunity to talk about their pets and how they act. Try to keep the focus primarily on dogs and cats for this lesson. After a cat owner describes the behavior of its cat, ask the dog owners if that is something their dog would do. Print the word *domesticated* on the chalkboard while explaining that these pets are referred to as domesticated animals because they rely upon humans for survival.

- **Pantomime Game:** Give each student two index cards on which to print the words *dog* and *cat*. Then give each child a Post-it Note on which to write a characteristic of a dog or cat, including those characteristics that demonstrate that it has been domesticated by humans. For example, *A cat will step into its litter box. Dogs can be taught to do tricks, like fetching a stick. A dog will come when you call him by name. Dogs and cats rely upon their owners to feed them. Dogs are bathed by humans.* Fold the papers over and put them in a lunch sack. Call on volunteers to draw a paper out of the sack and to act out the situation printed on it. After the other students have identified the action, they will hold up their "dog" or "cat" cards, depending on whether the behavior pantomimed was that of a dog or cat.
(<u>Supplies:</u> Index cards, Post-it Notes, lunch bag)

- **Writing/Handwriting assignment**
<u>Younger Students:</u> Position the handwriting assignment on page 217 under a Smart Document Camera or a similar device. Discuss the correct stroke for each letter as you copy the sentence onto your paper. Students should follow along with you, copying the letters on papers of their own. (This will help them to remain on task and to finish the handwriting assignment in unison.)

The incomplete sentence below can also be used for a writing assignment. Students can finish the sentence, add additional sentences, and include appropriate illustrations.

I think a _____ makes a good pet because...

<u>Older Students:</u> Older students can also write about the topic below. Let them copy the topic and respond to it in their writing journals. They may also want to add appropriate illustrations.

Write about a pet you have or would like to have.

Context Clues

Directions: Fill in each blank with a spelling word that makes sense in the context of the sentence. (Cross out the words from the list as you use them.) Circle the letters in the key words that make the short *e* sound (*e*, *ea*).

Spelling words: pet, bed, beg, fed, fetch, vet, smell, help, rest, sell, met, head

1. I have a ____________ dog named Hans, and I take good care of him.
2. I bathe him so he doesn't ____________.
3. I also play ____________ and other games with him.
4. When he gets tired, he has his very own ____________ to go to.
5. We got the bed at the pet store where they ____________ things for animals.
6. Mom told me that dogs need a lot of ____________.
7. I always make sure my dog is ____________.
8. Mom doesn't want me to give him people food from the table though because she said it teaches him to ____________.
9. When he gets sick, we take him to the ____________.
10. The vet knows how to ____________ him get better.
11. My dog loves me and has ____________ me after school every day since I got him.
12. I want him to know I love him, too, so I always pet him and scratch his ____________.

Spelling Words (Short *e* /ĕ/ sound)

K	C	N	V	B	N	S	E	L	L	U	O
H	S	R	E	S	T	S	U	T	F	B	H
E	U	O	D	R	A	E	M	O	T	J	E
L	T	P	M	V	E	T	V	D	X	H	A
P	P	E	C	H	N	R	D	J	Z	D	D
B	I	T	F	J	D	B	E	G	M	N	L
S	D	A	Q	V	K	M	W	C	Z	S	J
B	E	D	Z	F	F	H	F	K	G	T	E
G	S	L	W	E	G	H	E	C	R	W	M
X	G	Y	Y	T	M	E	D	H	I	Q	E
I	R	Q	E	C	P	X	Y	V	F	M	T
Q	Y	L	F	H	Z	S	M	E	L	L	X

Pets

1. pet
2. bed
3. beg
4. fed
5. fetch
6. vet
7. smell
8. help
9. rest
10. sell
11. met
12. head

Weekly Oral Language

1. Correct the asking sentence.

   may i walk the dog

2. Correct the telling sentence.

   you may walk the dog after breakfast

3. What is the subject of the telling sentence?

   ____

4. Read the *telling sentence* again and identify the helping verb and the verb (action word).

   Helping Verb: ______________________

   Verb: ______________________________

5. Write your own sentence about this week's theme.

   ____

Instruction: *Teaching Children about Capitalization, Punctuation, Apostrophe s, and Contractions (Language Book 2)* by Habakkuk Educational Materials can be purchased separately to give students the knowledge they need to complete this worksheet. After students have completed the worksheet and turned it in, complete a copy of it on your classroom screen.

Parts of Speech

Directions: Identify what parts of speech are used in the sentence below and write them in the spaces provided. Then identify the subject (naming part) and predicate (action part) of the sentence and write them in the spaces provided.

Parts of speech: Noun marker, noun, pronoun, verb, helping verb, adjective, adverb, preposition, conjunction, interjection

My	**dog**	**can**	**fetch**	**sticks.**
__________	__________	__________	__________	__________

Subject (Naming Part)

____

Predicate (Action Part)

____

Instruction: *Teaching Children about Parts of Speech (Language Book 3)* by Habakkuk Educational Materials can be purchased separately to help students quickly learn the parts of speech. After students have completed this worksheet and turned it in, complete a copy of it on your classroom screen.

Handwriting directions: Color the picture. Then copy the sentence below on the lines at the bottom of the page.

Pets and other animals that depend on humans for what they need are called domesticated.

Name: ______________________________

Addition and Subtraction Facts

Directions

1. Count the dogs above the first space and write the addend.
2. Count the dogs above the second space and write the addend.
3. Count all the dogs and write the sum after the equal sign.

______ + ______ = ______

Directions

1. Count the dogs above the first space and write the number.
2. Cross out 4 dogs.
3. Count the number of dogs left and write the answer after the equal sign.

______ − 4 = ______

Addition and Subtraction Word Problems

Directions: Color and cut out the puppies at the bottom of this page. Read each sentence and use the cutouts to help solve the problem. Write the addition or subtraction fact in the spaces provided.

1. My dog had **6** puppies. We sold **4** puppies. How many puppies do we have left?

   ___________ = _____

2. I had **2** cocker spaniels. The female had **4** puppies. How many spaniels do I have now?

   ___________ = _____

3. I had **1** cocker spaniel. She had **5** puppies. How many spaniels do I have now?

   ___________ = _____

4. My dog had **5** puppies. We gave **1** puppy away. How many puppies do we have left?

   ___________ = _____

QUESTIONS FOR THE THEMATIC UNITS GAME

Please refer to pages 28-29 of this book for instructions on how to use the unit questions.

What is a domesticated animal? *Lesson 11 (Science Themes)*	What is an example of a domesticated animal? *Lesson 11 (Science Themes)*

LESSON 11 TEST (Pets, Domesticated Animals)

1. What is a domesticated animal?

   ○ An animal that relies upon humans for survival

   ○ An animal that lives in the wild

   ○ An animal that lives in the ocean

2. What are examples of domesticated animals? (Mark TWO answers.)

   ○ An elephant that lives in the rainforest

   ○ Farm animals that rely upon humans for food

   ○ A pet like a dog or cat

Note: These cards can be used for the spelling word games listed on pages 21-22.

pet	bed
beg	fed
fetch	vet

Note: These cards can be used for the spelling word games listed on pages 21-22.

sm<u>e</u>ll	h<u>e</u>lp
r<u>e</u>st	s<u>e</u>ll
m<u>e</u>t	h<u>ea</u>d

Lesson 12

Theme: Farm Life
Sound: /or/
Spelling Patterns: *or*, *ore*

Spelling Words	Alphabetized
1. horse	____________
2. morning	____________
3. fork	____________
4. porch	____________
5. storm	____________
6. pork	____________
7. born	____________
8. corn	____________
9. orange	____________
10. store	____________
11. sore	____________
12. chores	____________

Inflectional endings: storms, storming, stormed

Bonus compound words: corncobs, pitchfork, porkchops, storeroom, thunderstorm, orange juice

Extra challenge word: orchard

Thematic Activities for Lessons 12 and 13
(Farm Life)

Thematic objectives: To learn about farm food and animals and the work of farmers

Key Questions for Discussion

- What do you call a calf that is a year old? (*a yearling*)
- When are most farm animals born? (*in the spring*)
- What makes bread rise? (*yeast*)
- Where do eggs come from that we eat for breakfast? (*hens, which are female chickens*)
- What do farmers do? (*Farmers plant and harvest foods to sell. They also raise animals.*)

Activities, Supplies, and Preparations for this Lesson

- **Field trip:** Lessons 12 and 13 have to do with farm life. Try to arrange a field trip to a local farm during this time.
  (Supplies: Signed permission slips for each child to attend a field trip to a local farm)

- **Spelling Patterns of Sounds Heard in the English Language:** Display words with this week's sound on the wall display. (See pages 26-27 for details.)

- **Reproducible materials:** Prepare the reproducible materials provided with this lesson, including the question cards and the spelling word cards
  (Supplies: Computer paper, cardstock)

- **Books and videos:** Read books and/or watch YouTube videos about farm life.

- **Phonics Store Game:** Follow the instructions on page 242.
  (Supplies: A lunch bag or a shoebox with a slit in the lid)

- **Writing/Handwriting assignment**

  Younger Students: Position the handwriting assignment on page 236 under a Smart Document Camera or a similar device. Discuss the correct stroke for each letter as you copy the sentence onto your paper. Students should follow along with you, copying the letters on papers of their own. (This will help them to remain on task and to finish the handwriting assignment in unison.)

  The incomplete sentences below can also be used for a writing assignment. Students can finish the sentences, add additional sentences, and include appropriate illustrations.

  The best part about our visit to the farm was...
  My favorite farm animal is...

  Older Students: Older students can write about the topic below. Let them copy the question and respond to it in their writing journals. They may also want to add appropriate illustrations.

  What did you like best about our visit to the farm?
  What is your favorite farm animal and why?

Context Clues

Directions: Fill in each blank with a spelling word that makes sense in the context of the sentence. (Cross out the words from the list as you use them.) Circle the letters in the key words that make the /or/ sound (*or*, *ore*).

Spelling words: horse, morning, fork, porch, storm, pork, born, corn, orange, store, sore, chores

1. Farmer Jim and his wife get up early every ____________.

2. They have many ____________ to do.

3. First, they will eat breakfast and drink their ____________ juice.

4. After they have eaten, they will use a ____________ to pitch hay into the manger for their animals to eat.

5. They have pigs and lots of other animals, but they will have even more in spring once the babies are ____________.

6. ____________, like ham and bacon, comes from pigs.

7. The couple also needs to pluck the ____________ today.

8. Later, Farmer Jim will brush his ____________ while his wife prepares dinner.

9. He will also gather the eggs that the mother chickens, called *hens*, have laid. They sell these in a ____________.

10. By the end of the day, their arms are very ____________.

11. After dinner this evening, they sat on the ____________ swing and watched the lightning in the distance.

12. Before the ____________ arrived, however, they went to bed for a good night's sleep.

Spelling Patterns of Sounds Heard in Our Spelling Words

Directions: Fill in the missing letters that make the /or/ sound.

- Use a **blue** colored pencil for ***or***.
- Use a **red** colored pencil for ***ore***.

h _ _ s e	s t _ _ m	_ _ a n g e
m _ _ n i n g	p _ _ k	s t _ _ _
f _ _ k	b _ _ n	s _ _ _
p _ _ c h	c _ _ n	c h _ _ _ s

Directions: Record spelling words whose /or/ sound is spelled *or* under the **"or"** headings and words whose /or/ sound is spelled *ore* under the **"ore" heading**.

or	or	ore
____________	____________	____________
____________	____________	____________
____________	____________	____________
____________	____________	____________
____________	____________	____________

Spelling Words (/or/ sound)

K	C	N	V	B	N	S	T	O	R	E	O
U	H	L	S	L	K	S	U	T	F	B	O
G	O	O	T	M	P	R	O	E	S	J	R
C	R	B	O	H	O	Y	V	D	O	H	A
K	E	K	R	H	R	R	B	J	R	D	N
B	S	N	M	J	C	O	O	L	E	N	G
S	D	A	Q	V	H	M	R	C	Z	S	E
H	O	R	S	E	F	H	N	K	G	T	E
G	S	L	W	E	G	H	B	C	O	R	N
X	P	O	R	K	M	E	P	H	I	Q	C
I	R	Q	E	I	P	X	Y	V	F	M	R
M	O	R	N	I	N	G	I	F	O	R	K

Farm Life Words

1. horse
2. morning
3. fork
4. porch
5. storm
6. pork
7. born
8. corn
9. orange
10. store
11. sore
12. chores

Homophones

horse	**hoarse**

My **horse** is walking around the pasture.	My voice is **hoarse**.
A horse is an animal.	*Hoarse refers to a temporary change in a person's voice.*

My **horse** can't hear me calling because my voice is **hoarse**.

Underline the word that refers to an animal.
Circle the word that refers to a temporary change in a person's voice.

Homophones

morning | mourning

Our rooster crows every **morning**.

Morning is the early part of the day.

The kid has been **mourning** ever since he lost his dog.

Mourning means "to feel or show sadness."

The boy is **mourning** because he lost his dog this **morning**.

Underline the word that refers to the early part of the day.
Circle the word that means "to feel or show sadness."

Homophones

soar

Watch the eagle **soar** through the sky.

To soar means "to fly or glide through the air."

sore

I am **sore** all over!

Sore means "very painful."

My neck is **sore** from staring up to watch the eagle **soar**.

Underline the word that means "to fly or glide through the air."
Circle the word that means "very painful."

Compound Words

Directions: When two or more root words are used together to form one word, it is called a **compound word** (homemade). Form six compound words by adding a spelling word to the words *pitch*, *thunder*, *cobs*, *chops*, *room*, and *juice*.

Spelling words: corn, fork, orange, pork, store, storm

pitch_______ thunder_______ _______cobs

_______chops _______room _______ juice

Directions: Fill in each blank with one of the compound words that makes sense in the context of the sentence.

This morning, the farmer ate breakfast and drank homemade _______________ that his wife made. Then he went outside and used a _______________ to pitch hay into the manger for his farm animals to eat. They worked together to pluck the _______________ from the stalk and to store it in their _______________. While relaxing at the dinner table, they ate _______________ and watched the _______________ from the kitchen window.

Weekly Oral Language

1. Correct the asking sentence.

   when are most farm animals born

2. Correct the telling sentence.

   most farm animals are born in the spring

3. What is the subject of the telling sentence?

   ____

4. Read the *telling sentence* again and identify the helping verb and the verb (action word).

   Helping Verb: ______________________

   Verb: ______________________

5. Write your own sentence about this week's theme.

   ____

Instruction: *Teaching Children about Capitalization, Punctuation, Apostrophe s, and Contractions (Language Book 2)* by Habakkuk Educational Materials can be purchased separately to give students the knowledge they need to complete this worksheet. After students have completed the worksheet and turned it in, complete a copy of it on your classroom screen.

Parts of Speech

Directions: Identify what parts of speech are used in the sentence below and write them in the spaces provided. Then identify the subject (naming part) and predicate (action part) of the sentence and write them in the spaces provided.

Parts of speech: Noun marker, noun, pronoun, verb, helping verb, adjective, adverb, preposition, conjunction, interjection

A	farmer	has	many	chores.
________	________	________	________	________

Subject (Naming Part)

____

Predicate (Action Part)

____

Instruction: *Teaching Children about Parts of Speech (Language Book 3)* by Habakkuk Educational Materials can be purchased separately to help students quickly learn the parts of speech. After students have completed this worksheet and turned it in, complete a copy of it on your classroom screen.

Handwriting directions: Position the page under a Smart Document Camera or a similar device. Discuss the correct stroke for each letter as you copy the sentences onto your paper. Students should follow along with you, copying the letters on papers of their own. (This will help them to remain on task and to finish the handwriting assignment in unison.) Allow time for students to color their pictures.

We call a mother chicken a hen. The eggs we eat for breakfast are laid by hens.

Name: ______________________

Addition and Subtraction Facts

Directions
1. Count the barns above the first space and write the addend.
2. Count the barns above the second space and write the addend.
3. Count all the barns and write the sum after the equal sign.

______ + ______ = ______

Directions
1. Count the barns above the first space and write the number.
2. Cross out 3 barns.
3. Count the number of barns left and write the answer after the equal sign.

______ − 3 = ______

QUESTIONS FOR THE THEMATIC UNITS GAME

Please refer to pages 28-29 of this book for instructions on how to use the unit questions.

What do you call a calf that is a year old? *Lessons 12 and 13 (Social Studies)*	When are most farm animals born? *Lessons 12 and 13 (Social Studies)*
What makes bread rise? *Lessons 12 and 13 (Social Studies)*	Where do eggs come from that we eat for breakfast? *Lessons 12 and 13 (Social Studies)*
What do farmers do? *Lessons 12 and 13 (Social Studies)*	

LESSONS 12-13 TEST (Farm Life)

1. What do you call a calf that is a year old?
   - ◯ Yearling
   - ◯ Kid
   - ◯ Cub

2. When are most farm animals born?
   - ◯ Winter
   - ◯ Spring
   - ◯ Fall

3. What makes bread rise?
   - ◯ Dough
   - ◯ Salt
   - ◯ Yeast

4. Where do eggs come from that we eat for breakfast?
   - ◯ Hens, which are female chickens
   - ◯ Eagles
   - ◯ Doves

5. What do farmers do?
   - ◯ Plant and harvest foods to sell
   - ◯ Raise animals
   - ◯ Both of the above answers are true.

Note: These cards can be used for the spelling word games listed on pages 21-22.

h<u>or</u>se	m<u>or</u>ning
f<u>or</u>k	p<u>or</u>ch
st<u>or</u>m	p<u>or</u>k

Note: These cards can be used for the spelling word games listed on pages 21-22.

born	corn
orange	store
sore	chores

Phonics Store Game: Cut out the store below and tape it to a lunch bag. Cut out the spelling word cards as well as the word cards having *or* that makes the /ər/ sound as in *col<u>or</u>*. Each child will choose a card to read and note if the word makes the /or/ or /ər/ sound. If it makes the /or/ sound heard in *store*, the child will stick it in the bag. (Word cards representing both the /or/ and /ər/ sounds are provided, but only those with the /or/ sound will be placed in the bag.)

Rule: The letter combinations *ar* and *or* make the /ər/ sound when they end a word with more than one syllable. (Mol<u>ar</u>, col<u>or</u>)

Note: Flashcards on this page contain *or* pronounced as /ər/ rather than /or/ and are used only with the "Store Game" described on the previous page.

author	color
doctor	mayor
pastor	sailor

Lesson 13

Theme: Farm Life
Sound: /y/
Spelling Pattern: *y*

Spelling Words

1. young
2. yams
3. yak
4. yell
5. yoke
6. yolk
7. yearling
8. yarn
9. yonder
10. yeast
11. yellow
12. farmyard

Alphabetized

____
____
____
____
____
____
____
____
____
____
____
____

Inflectional endings: yells, yelling, yelled; younger, youngest

Bonus compound word: barnyard

Extra challenge word: onion

Context Clues

Directions: Fill in each blank with a spelling word that makes sense in the context of the sentence. (Cross out the words from the list as you use them.) Circle the letter in the key words that make the /y/ sound.

Spelling words: young, yams, yak, yell, yoke, yolk, yearling, yarn, yonder, yeast, yellow, farmyard

1. I'm visiting my grandparents, and they have a ____________ full of animals.
2. They have cows, which are very protective of their ____________.
3. One of the calves is a year old, so it is referred to as a ____________.
4. They even have a ____________!
5. I especially like their cute, ____________ chicks.
6. Their full-grown chickens sometimes lay double ____________ eggs.
7. Today, Grandma is going to show me how to grow ____________.
8. She is also going to teach me how to knit with ____________.
9. Grandpa will be busy plowing down ____________.
10. He attaches the ____________ to the necks of their horses, and they do a lot of the work.
11. Grandma is making homemade bread for dinner, and she explained to me that something called ____________ makes the bread rise.
12. When dinner is ready, I will go outside to ____________ for Grandpa, and then we all sit down to eat.

Spelling Words (/y/ sound)

Y	C	Y	E	A	S	T	N	E	Y	U	Y
E	K	L	I	L	K	S	U	T	O	B	E
L	U	Y	E	L	L	O	W	A	K	J	A
L	T	B	M	H	T	Y	V	D	E	H	R
K	Y	O	U	N	G	R	Y	J	Z	D	L
B	I	N	F	J	D	O	O	L	Y	N	I
S	Y	A	M	S	K	M	N	C	O	S	N
A	F	N	Z	Q	F	H	D	K	L	T	G
Y	S	Y	A	R	N	H	E	C	K	W	O
A	G	P	K	C	H	N	R	H	I	Q	C
K	R	Q	E	I	P	X	Y	L	W	E	G
Q	F	A	R	M	Y	A	R	D	P	J	X

Farm Life Words

1. young
2. yams
3. yak
4. yell
5. yoke
6. yolk
7. yearling
8. yarn
9. yonder
10. yeast
11. yellow
12. farmyard

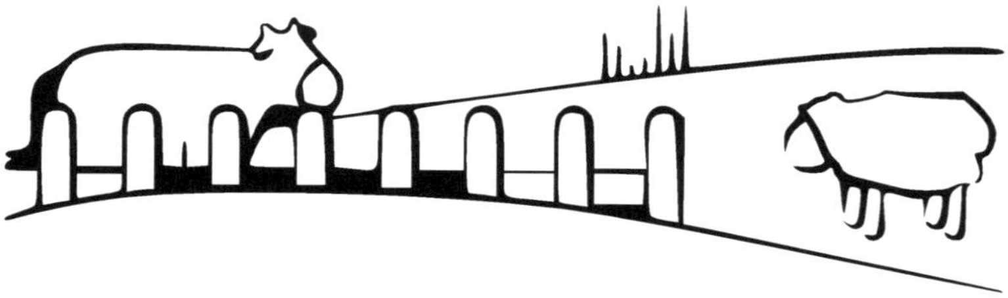

Homophones

yolk

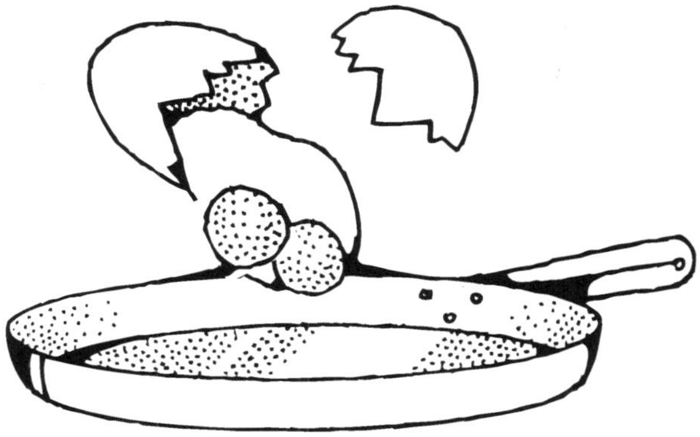

Dad likes to dip his toast in the **yolk** of his fried eggs.

Yolk is the yellow part of an egg.

yoke

The **yoke** allows the horses to pull the plow.

A yoke is a wooden device that is attached to the necks of two animals and to the plow they pull.

The farmer's wife is frying him a double **yolk** egg while he removes the **yoke** from the horses.

Underline the word that refers to the yellow part of an egg.
Circle the word that refers to a wooden device that is attached to the necks of two animals and to the plow they pull.

Weekly Oral Language

1. Correct the asking sentence.

   what is a year-old calf called

2. Correct the telling sentence.

   a year-old calf is called a yearling

3. What is the subject of the telling sentence?

   ____

4. Read the *telling sentence* again and identify the helping verb and the verb (action word).

   Helping Verb: ______________________

   Verb: ______________________________

5. Write your own sentence about this week's theme.

   ____

Instruction: *Teaching Children about Capitalization, Punctuation, Apostrophe s, and Contractions (Language Book 2)* by Habakkuk Educational Materials can be purchased separately to give students the knowledge they need to complete this worksheet. After students have completed the worksheet and turned it in, complete a copy of it on your classroom screen.

Parts of Speech

Directions: Identify what parts of speech are used in the sentence below and write them in the spaces provided. Then identify the subject (naming part) and predicate (action part) of the sentence and write them in the spaces provided.

Parts of speech: Noun marker, noun, pronoun, verb, helping verb, adjective, adverb, preposition, conjunction, interjection

Yeast	makes	the	bread	rise.
__________	__________	__________	__________	__________

Subject (Naming Part)

____

Predicate (Action Part)

____

Instruction: *Teaching Children about Parts of Speech (Language Book 3)* by Habakkuk Educational Materials can be purchased separately to help students quickly learn the parts of speech. After students have completed this worksheet and turned it in, complete a copy of it on your classroom screen.

Note: These cards can be used for the spelling word games listed on pages 21-22.

young	yams
yak	yell
yoke	yolk

Note: These cards can be used for the spelling word games listed on pages 21-22.

yearling	yarn
yonder	yeast
yellow	farmyard

Lesson 14

Theme: Thanksgiving in America
Sound: /th/ in *thank* and /th/ in *Father*
Spelling Pattern: *th*

Spelling Words	Alphabetized
1. Thanksgiving	______________
2. thank	______________
3. thankful	______________
4. Thursday	______________
5. Plymouth	______________
6. month	______________
7. feathers	______________
8. together	______________
9. gather	______________
10. Father	______________
11. them	______________
12. they	______________

Inflectional endings: thanks, thanking, thanked

Bonus compound word: themselves

Extra challenge word: fourth

Note: The initial letters of *Thanksgiving*, *Thursday*, and *Plymouth* must be capitalized.

Thematic Activities for Lesson 14
(Thanksgiving in America)

Thematic objectives: To acquaint students with events that led to the first Thanksgiving Day and to help instill in them a sense of thankfulness

Key Questions for Discussion

- Thanksgiving is a special day to give thanks to our heavenly Father for all His blessings. What are some things you are thankful for?
- When is Thanksgiving celebrated in America? (*the fourth Thursday of November*)
- The history of the first Thanksgiving began in 1620 when the Pilgrims left England on a big ship and came to America. What was the name of the ship they sailed on to the New World? (*the Mayflower*)
- After the Pilgrims settled in the New World, which is America, what did they call their new home? (*Plymouth Colony*)
- What state is Plymouth Colony in today? (*Massachusetts*)
- What was the name of the Indian who knew the English language and was able to teach the Pilgrims how to survive in their new home? (*Squanto*)
- The Pilgrims built houses to live in and planted crops to eat. After their first harvest in 1621, who celebrated the first Thanksgiving together? (*the Pilgrims and Indians*)
- What were the Pilgrims thankful for? (*their new country, their homes, and their food*)
- In 1863, what president declared that Thanksgiving would be a federal holiday to be celebrated every November? (*Abraham Lincoln*)
- How do families today celebrate Thanksgiving? (*Families get together to enjoy a big Thanksgiving meal with turkey, pumpkin pie, and lots of side dishes.*)

Activities, Supplies, and Preparations for this Lesson

- **Spelling Patterns of Sounds Heard in the English Language:** Display words with this week's sound on the wall display. (See pages 26-27 for details.)

- **Reproducible materials:** Prepare the reproducible materials provided with this lesson, including the question cards and the spelling word cards
  (Supplies: Computer paper, cardstock)

- **Books and videos:** Read books and/or watch YouTube videos about the first Thanksgiving. (For any books you read, identify the title, the author, and the illustrator. Also, examine the front and back cover of the book before making any predictions about the story or discussing what it is about.)
  (Suggested YouTube Video: *The Story of the First Thanksgiving*, by Elaine Raphael and Don Bolognese, video by Bookworm Bonny)

- **Indian Headdresses:** Purchase a bag of feathers from Walmart or somewhere else. Explain to the children that Indians had to earn their feathers for different acts of bravery. Throughout the week, let students earn feathers for solving math problems correctly or for any other accomplishment of the teacher's choice. At the end of the week, use the feathers to make Indian headdresses. Try to make sure everyone earns an equal number of feathers.

  Instructions: Fold a piece of construction paper in half longways. Then take one half and fold it in half to the middle crease. Do the same thing with the opposite half. When you cut down the middle crease, you will have a band for two children. Fold a band and punch a hole at both ends. Tape feathers inside the crease and then glue the open end shut. Tie a string of yarn through the hole on each end and tie to the child's head.
  (<u>Supplies:</u> A bag of feathers purchased from Walmart or elsewhere; construction paper; hole punch; tape; Elmer's glue; yarn)

- **Thanksgiving Booklets:** On page 262, you will find four strips with sentences relating to the first Thanksgiving celebration. Discuss with students why one event could not have happened before or after another. Instruct students to cut out the strips and to put them in the correct order. When they have finished, let them paste one on each page of their booklets, being careful that the event that occurred first is on page 1, the second event to occur is on page 2, and so forth. Students will add appropriate illustrations, tell who the booklet was illustrated by on the cover page, and record a title (e.g., "The First Thanksgiving").

  Instructions for assembling the booklets for children: Put a white sheet of construction paper atop a colored sheet of construction paper (light fall colors like yellow and orange are recommended). Fold them in half so that the colored paper, which will be the book cover, is on the outside and four white pages are inside. Fasten the papers together at the crease with brads.
  (<u>Supplies:</u> Orange or yellow and white construction paper, paste, hole punch, brads)

- **Geography:** Use a world map to show students where the Pilgrims set sail from in the Mayflower (England) and where they eventually arrived at in the New World (Massachusetts).

- **Writing/Handwriting assignment:** Position the handwriting assignment included with this lesson (i.e., the page with the family enjoying a Thanksgiving meal together) under a Smart Document Camera or a similar device. Discuss the correct stroke for each letter as you copy the incomplete sentence onto your paper. Students should follow along with you, copying the letters on papers of their own. (This will help them to remain on task and to finish the handwriting assignment in unison.) They can then finish the sentence and add additional sentences.

Context Clues

Directions: Fill in each blank with a spelling word that makes sense in the context of the sentence. (Cross out the words from the list as you use them.) Circle the letters in the key words that make the /th/ sound in *thank* and the /th/ sound in *Father*.

Spelling words: Thanksgiving, thank, thankful, Thursday, Plymouth, month, feathers, together, gather, Father, them, they

1. Thanksgiving is a day to give thanks to our heavenly __________.

2. In America, Thanksgiving is celebrated during the __________ of November.

3. It is always on the fourth __________ of the month.

4. The history of the first _______________, however, began in 1620, when the Pilgrims left England on a big ship called the Mayflower.

5. After many days of sailing, they arrived in the New World, which is America, and called their new home __________ Colony. If you want to visit Plymouth today, you will need to go to Massachusetts.

6. Winter was arriving, and __________ worked hard to build homes.

7. Soon, they met some new friends who wore __________ in their hair. They were Indians, and one of them spoke English. His name was Squanto.

8. The Pilgrims were __________ for Squanto because he did a lot to help them know how to survive in their new home.

9. One of the things Squanto taught __________ was how to grow corn.

10. In 1621, after their first harvest, the Indians and the Pilgrims celebrated the first Thanksgiving __________.

11. Today, families still __________ together to enjoy a big Thanksgiving meal with turkey and lots of side dishes.

12. We pray to the Father and __________ Him for the food and for all His blessings.

Spelling Patterns of Sounds Heard in Our Spelling Words

Directions: Fill in the missing letters.

- Use a **blue** colored pencil for the /th/ sound heard in *thank*.
- Use a **red** colored pencil for the /th/ sound heard in *Father*.

_ _ a n k s g i v i n g	f e a _ _ e r s
_ _ a n k	t o g e _ _ e r
_ _ a n k f u l	g a _ _ e r
_ _ u r s d a y	F a _ _ e r
P l y m o u _ _	_ _ e m
m o n _ _	_ _ e y

Directions: Categorize the spelling words. If the *th* in a word makes the sound of *th* in *thank*, record it under the "*th* in *thank*" heading. If it makes the sound of *th* in *Father*, record it under the "*th* in *Father*" heading.

th* in *thank	

th* in *Father	

Spelling Words (/th/ and /th/ sound)

K	C	F	A	T	H	E	R	E	G	M	T
U	S	F	I	L	K	S	U	T	L	O	H
T	U	T	H	U	R	S	D	A	Y	N	A
H	T	B	M	H	T	Y	V	D	X	T	N
A	P	L	Y	M	O	U	T	H	F	H	K
N	I	T	F	J	D	O	A	L	E	N	S
K	A	H	T	H	A	N	K	C	A	S	G
F	F	E	Z	Q	T	H	E	Y	T	T	I
U	S	M	W	E	G	H	B	C	H	W	V
L	G	G	A	T	H	E	R	H	E	Q	I
I	R	Q	E	I	P	X	Y	V	R	M	N
T	O	G	E	T	H	E	R	J	S	J	G

Thanksgiving Words

1. Thanksgiving
2. thank
3. thankful
4. Thursday
5. Plymouth
6. month
7. feathers
8. together
9. gather
10. Father
11. them
12. they

Directions: Add the prefix and suffixes listed below to the root word *thank* to form a word that means "full of thanks," "not thankful," and "being full of thanks."

Prefix: *un-*
Suffixes: -*ful*, -*ness*

Prefix	**Root Word**	**Suffix**	
__________	thank	__________	__________
__________	thank	__________	__________
__________	thank	__________	__________

Which word means "full of thanks"? ____________________

Which word means "not thankful"? ____________________

Which word is a noun meaning "being full of thanks"? __________

Directions: Insert a prefix or suffix in the space provided to form a new word that makes sense in the context of the sentence.

Thanksgiving is a special time to **thank** God and to remind us to be **thank**_____ for the good things He provides. We must never be _____**thank**_____, so every day, not just on Thanksgiving, we should show are **thank**__________.

- A **prefix** is added to the beginning of a root word to form a new word (<u>al</u>together). A **suffix** is added to the end of a root word to form a new word (thank<u>ful</u>). Both prefixes and suffixes are **affixes**.
- **Derivative:** A root word with something added to it (e.g., a prefix or a suffix)
- **Rule:** When a consonant suffix is added to a word, just put the word and the suffix together. (Thank<u>ful</u>)

Weekly Oral Language

1. Correct the asking sentence.

   what ship did the Pilgrims sail on to america

2. Correct the telling sentence.

   the Pilgrims sailed on the Mayflower to america

3. What is the subject of the telling sentence?

   ____

4. Read the *telling sentence* again and identify the verb (action word).

   ____

5. Write your own sentence about this week's theme.

   ____

Instruction: *Teaching Children about Capitalization, Punctuation, Apostrophe s, and Contractions (Language Book 2)* by Habakkuk Educational Materials can be purchased separately to give students the knowledge they need to complete this worksheet. After students have completed the worksheet and turned it in, complete a copy of it on your classroom screen.

Parts of Speech

Directions: Identify what parts of speech are used in the sentence below and write them in the spaces provided. Then identify the subject (naming part) and predicate (action part) of the sentence and write them in the spaces provided.

Parts of speech: Noun marker, noun, pronoun, verb, helping verb, adjective, adverb, preposition, conjunction, interjection

The	**Pilgrims**	**founded**	**Plymouth**	**Colony.**
__________	__________	__________	__________	__________

Subject (Naming Part)

____

Predicate (Action Part)

____

Instruction: *Teaching Children about Parts of Speech (Language Book 3)* by Habakkuk Educational Materials can be purchased separately to help students quickly learn the parts of speech. After students have completed this worksheet and turned it in, complete a copy of it on your classroom screen.

Strips for the Thanksgiving Booklets

After their first harvest in Plymouth Colony, the Pilgrims and Indians celebrated the first Thanksgiving together.

In 1620, the Pilgrims left England and sailed to America on a big ship called the Mayflower.

Years later in 1863, President Abraham Lincoln declared that Thanksgiving would be a federal holiday to be celebrated every November.

When they reached America, the Pilgrims built houses to live in and planted gardens.

Handwriting directions: Color the picture. Then copy "I am thankful for" on the lines at the bottom of the page. Complete the sentence by telling what you are thankful for.

I am thankful for...

Name: ______________________

Addition and Subtraction Facts

Directions

1. Count the turkeys above the first space and write the addend.
2. Count the turkeys above the second space and write the addend.
3. Count all the turkeys and write the sum after the equal sign.

________ + ________ = ________

Directions

1. Count the turkeys above the first space and write the number.
2. Cross out 2 turkeys.
3. Count the number of turkeys left and write the answer after the equal sign.

________ − 2 = ________

Subtraction Word Problems

Directions: Color and cut out the turkeys at the bottom of the page. Read each sentence and use the cutouts to help solve the problem. Write the subtraction fact in the spaces provided.

1. There were 18 turkeys. Nine were eaten on Thanksgiving. How many turkeys are left?

   _____ – _____ = _____

2. There were 15 turkeys. Seven were eaten on Thanksgiving. How many turkeys are left?

   _____ – _____ = _____

3. There were 13 turkeys. Eight were eaten on Thanksgiving. How many turkeys are left?

   _____ – _____ = _____

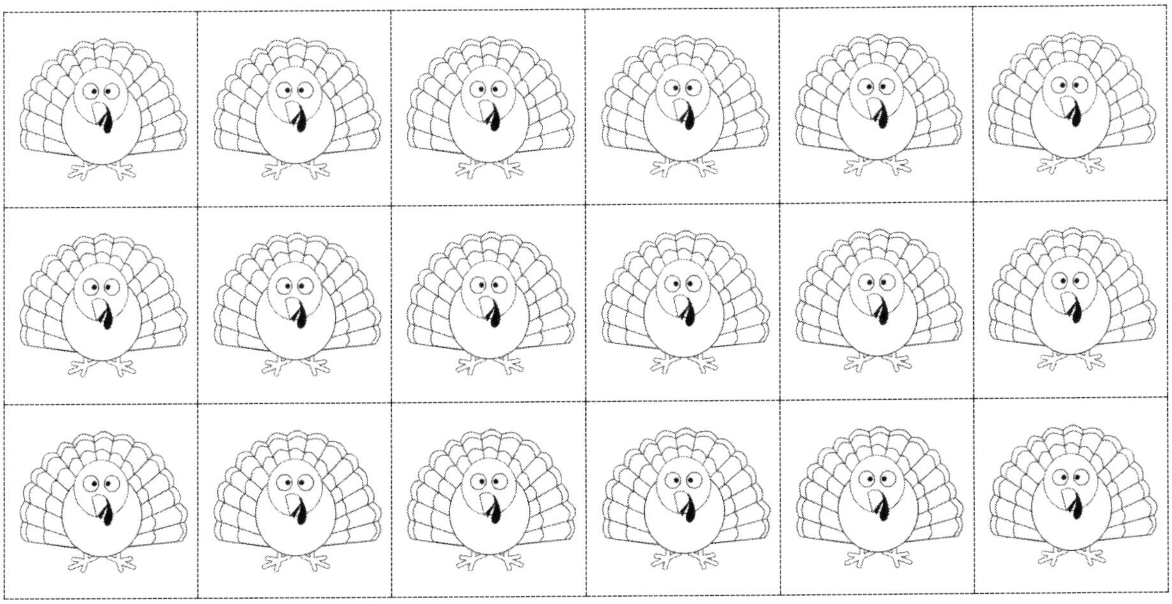

QUESTIONS FOR THE THEMATIC UNITS GAME

Please refer to pages 28-29 of this book for instructions on how to use the unit questions.

Thanksgiving is a special day to give thanks to our heavenly Father for all His blessings. What are some things you are thankful for? *Lesson 14 (Social Studies)*	When is Thanksgiving celebrated in America? *Lesson 14 (Social Studies)*
The history of the first Thanksgiving began in 1620 when the Pilgrims left England on a big ship and came to America. What was the name of the ship they sailed on to the New World? *Lesson 14 (Social Studies)*	After the Pilgrims settled in the New World, which is America, what did they call their new home? *Lesson 14 (Social Studies)*
Plymouth Colony is what the Pilgrims called their new home. What state is Plymouth Colony in today? *Lesson 14 (Social Studies)*	What was the name of the Indian who knew the English language and was able to teach the Pilgrims how to survive in their new home? *Lesson 14 (Social Studies)*
The Pilgrims built houses to live in and planted crops to eat. After their first harvest in 1621, who celebrated the first Thanksgiving together? *Lesson 14 (Social Studies)*	What were the Pilgrims thankful for? *Lesson 14 (Social Studies)*
In 1863, what president declared that Thanksgiving would be a federal holiday to be celebrated every November? *Lesson 14 (Social Studies)*	How do families today celebrate Thanksgiving? *Lesson 14 (Social Studies)*

LESSON 14 TEST (Thanksgiving in America)

1. What is the true purpose of Thanksgiving?

   ○ To eat turkey and pumpkin pie

   ○ To get a break from school and work

   ○ To give thanks to God for all His blessings

2. When is Thanksgiving celebrated in America?

   ○ October 31

   ○ The fourth Thursday of November

   ○ December 25

3. The history of the first Thanksgiving began in 1620 when the Pilgrims left England on a big ship and came to America. What was the name of the ship they sailed on to the New World?

   ○ Maria

   ○ The Titanic

   ○ The Mayflower

4. After the Pilgrims settled in the New World, which is America, what did they call their new home?

   ○ Plymouth Colony

   ○ Massachusetts

   ○ The Mayflower

5. What state is Plymouth Colony in today?

- ◯ Florida
- ◯ Oklahoma
- ◯ Massachusetts

6. What was the name of the Indian who knew the English language and was able to teach the Pilgrims how to survive in their new home?

- ◯ Squanto
- ◯ Sitting Bull
- ◯ Geronimo

7. The Pilgrims built houses to live in and planted crops to eat. After their first harvest in 1621, who celebrated the first Thanksgiving together?

- ◯ The Pilgrims and the Spanish
- ◯ The Pilgrims and Indians
- ◯ The Pilgrims and President Lincoln

8. What were the Pilgrims thankful for?

- ◯ Their new country
- ◯ Their homes and food
- ◯ Both of the above answers are true.

9. In 1863, what president declared that Thanksgiving would be a federal holiday to be celebrated every November?

- ◯ George Washington
- ◯ Abraham Lincoln
- ◯ Jimmy Carter

Note: These cards can be used for the spelling word games listed on pages 21-22.

Thanksgiving	thank
thankful	Thursday
Plymouth	month

Note: These cards can be used for the spelling word games listed on pages 21-22.

feathers	together
gather	Father
them	they

Lesson 15

Theme: Day and Night
Sound: /m/
Spelling Patterns: *m*, *me*

Spelling Words	Alphabetized
1. meals	____________
2. moon	____________
3. muffins	____________
4. morning	____________
5. midnight	____________
6. dream	____________
7. alarm	____________
8. pajamas	____________
9. home	____________
10. time	____________
11. daytime	____________
12. nighttime	____________

Inflectional endings: dreams, dreaming, dreamed

Bonus compound words: homework, oatmeal, alarm clock, full moon, half-moon

Thematic Activities for Lesson 15
(Day and Night)

Thematic objectives: To learn why we have day when it is light outside and night when it is dark

Key Questions for Discussion

- Why do we have day and night? (The earth rotates on its axis so that about half of the earth is always facing the sun while the other half is not. When the part of the earth you live on is facing the sun, you have daytime, but it is dark at the same time in another part of the world.)
- How long does it take for the earth to spin completely around? (*24 hours*)

Activities, Supplies, and Preparations for this Lesson

- **Spelling Patterns of Sounds Heard in the English Language:** Display words with this week's sound on the wall display. (See pages 26-27 for details.)
- **Reproducible materials:** Prepare the reproducible materials provided with this lesson, including the question cards and the spelling word cards
  (Supplies: Computer paper, cardstock)
- **Books and videos:** Read books and/or watch YouTube videos for kids about day and night. (For any books you read, identify the title, the author, and the illustrator. Also, examine the front and back cover of the book before making any predictions about the story or discussing what it is about.)
- **Day and Night Experiment:** Use a globe to show your students where you live. It would be helpful, especially for younger students, to put a sticker on the location just long enough for the experiment. Position a flashlight so that it's shining in the location of your country. Tell students that it represents the sun. Then slowly spin the globe until the flashlight is no longer shining on the sticker. Explain to students, *The earth rotates on its axis, and that's why we have day and night periods. The sun is shining where we live right now, but it is dark in another part of the world. The earth will continue to rotate today until we are no longer facing the sun. Then it will be dark where we live, and the sun will be shining in another part of the world. When half of the earth is dark, the other half is light.*

  Continue rotating the globe until the sticker is once again lit up by the flashlight. Tell students, *We sleep when it is dark, but it is light when we get up in the morning because the earth has spun around so that we are facing the sun again. It takes 24 hours (1 day and night) for the earth to spin completely around.*
  (Supplies: Globe, flashlight)

- **Crafts:** Use a white crayon to draw stars on a piece of white construction paper. (You will need to press down hard with the crayon.) The stars will become visible as students paint over them with black tempera thinned with water.
(Supplies: White crayon, white construction paper, black tempera thinned with water)
- **Writing/Handwriting assignment**
Younger Students: Make a copy of the reproducible writing paper provided on page 30 and position it under a Smart Document Camera or a similar device. Discuss the correct stroke for each letter as you copy the incomplete sentences below (or one of your own choosing) onto your paper. Students should follow along with you, copying the letters on papers of their own. (This will help them to remain on task and to finish the handwriting assignment in unison.) They can then finish the sentence, add additional sentences, and include appropriate illustrations.

  In the daytime, I enjoy _____.
  At nighttime, I enjoy _____.

  Older Students: Older students can write about the topic below. Let them copy the question and respond to it in their writing journals. They may also want to add appropriate illustrations.

  Do you like the day or the nighttime better and why?

- **A step-by-step process for using the Language-Experience Approach (LEA) to write a class story:**
  1. The teacher and her pupils enjoy a daytime experience together, such as a nature walk around the school neighborhood.
  2. Upon returning to the classroom, they discuss this experience, recalling the activities they observed while outdoors.
  3. The teacher records the mentioned activities on the chalkboard.
  4. A story can then be printed on chart paper and on poster board with the key words deleted. (See the illustration for an example.)

     Deleted words: sun, moon, birds, bikes, asleep

     While we toured the school neighborhood, we noticed that the _____ was shining bright. We looked for the _____, but it couldn't be found, for it only comes out at night. We heard _____ singing and saw people riding _____. They weren't _____, as I am at night.

  5. The deleted words can be printed on strips of paper, and the children can tape them in place while reading the story.
  6. They can then copy the story on papers of their own. They may also wish to glue magazine picture cutouts, etc. on their papers, showing what they observed on their walk.

Context Clues

Directions: Fill in each blank with a spelling word that makes sense in the context of the sentence. (Cross out the words from the list as you use them.) Circle the letters in the key words that make the /m/ sound (*m*, *me*).

Spelling words: meals, moon, muffins, morning, midnight, dream, alarm, pajamas, home, time, daytime, nighttime

1. The sun shines during the ______________.
2. At __________________, the stars can be seen.
3. The ______________ also shines at night.
4. I sometimes ______________ during the night.
5. Then my ______________ clock wakes me up for school.
6. I eat three ______________ every day.
7. I eat breakfast in the ______________, lunch in the afternoon, and dinner at night.
8. ______________ are my favorite breakfast food!
9. I go to school during the daytime, and I stay ______________ at night.
10. When it's ______________ for bed, I brush my teeth.
11. Then I put on my ______________ and tell my parents goodnight.
12. I almost always go to bed early, but once a year, on New Year's Eve, my parents let me stay up until ______________ to welcome in the new year!

Spelling Patterns of Sounds Heard in Our Spelling Words

Directions: Fill in the missing letters that make the /m/ sound.

- Use a **blue** colored pencil for ***m***.
- Use a **red** colored pencil for ***me***.

_ e a l s	_ i d n i g h t	h o _ _
_ o o n	d r e a _	t i _ _
_ u f f i n s	a l a r _	d a y t i _ _
_ o r n i n g	p a j a _ a s	n i g h t t i _ _

Directions: Record spelling words whose /m/ sound is spelled *m* under the **"m"** headings and words whose /m/ sound is spelled *me* under the **"me"** heading.

m	m	me
__________	__________	__________
__________	__________	__________
__________	__________	__________
__________	__________	__________

Spelling Words (/m/ sound)

K	C	M	I	D	N	I	G	H	T	U	N
U	S	L	I	L	K	M	E	A	L	S	I
M	O	R	N	I	N	G	O	E	M	J	G
C	T	B	M	H	T	Y	A	D	U	H	H
K	T	I	M	E	N	R	L	J	F	D	T
P	I	N	F	J	D	O	A	L	F	N	T
A	A	H	O	M	E	M	R	C	I	S	I
J	F	N	Z	Q	F	H	M	K	N	T	M
A	S	M	W	E	G	H	B	C	S	W	E
M	G	O	Y	D	R	E	A	M	I	Q	C
A	R	O	E	I	P	X	Y	V	F	M	R
S	Y	N	F	D	A	Y	T	I	M	E	X

Day and Night Words

1. meals
2. moon
3. muffins
4. morning
5. midnight
6. dream
7. alarm
8. pajamas
9. home
10. time
11. daytime
12. nighttime

Homophones

morning

Our rooster crows
every **morning**.

Morning is the early part of the day.

mourning

The kid has been **mourning**
ever since he lost his dog.

*Mourning means "to
feel or show sadness."*

The boy is **mourning** because
he lost his dog this **morning**.

Underline the word that refers to the early part of the day.
Circle the word that means "to feel or show sadness."

Compound Words

Directions: When two or more root words are used together to form one word, it is called a **compound word** (<u>some</u>where). Form seven compound words by adding a spelling word to the words *day*, *night*, *oat*, *work*, *half-*, and *clock*.

Spelling words: alarm, home, meal, mid, moon, time

day_______ night_______ oat_______ _______work

_______night half-_______ _______ clock

Directions: Fill in each blank with one of the compound words that makes sense in the context of the sentence.

My _______________ woke me up this morning. I ate some _______________ for breakfast and then went to school. Kids spend most of their _______________ at school. After school, I went home and played and did my _______________. I was ready for bed by _______________. As I put on my pajamas, I glanced out the window and noticed that there was a _______________ in the sky. A few hours later at _______________, I woke back up and glanced out my window again. This time I couldn't see the moon, but it was still somewhere in the sky!

Weekly Oral Language

1. Correct the asking sentence.

   what can we eat for breakfast

2. Correct the telling sentence.

   you can eat muffins for breakfast

3. What is the subject of the telling sentence?

   ____

4. Read the *telling sentence* again and identify the helping verb and the verb (action word).

   Helping Verb: ______________________

   Verb: ______________________________

5. Write your own sentence about this week's theme.

   ____

Instruction: *Teaching Children about Capitalization, Punctuation, Apostrophe s, and Contractions (Language Book 2)* by Habakkuk Educational Materials can be purchased separately to give students the knowledge they need to complete this worksheet. After students have completed the worksheet and turned it in, complete a copy of it on your classroom screen.

Parts of Speech

Directions: Identify what parts of speech are used in the sentence below and write them in the spaces provided. Then identify the subject (naming part) and predicate (action part) of the sentence and write them in the spaces provided.

Parts of speech: Noun marker, noun, pronoun, verb, helping verb, adjective, adverb, preposition, conjunction, interjection

The	moon	is	shining	brightly.
________	________	________	________	________

Subject (Naming Part)

____

Predicate (Action Part)

____

Instruction: *Teaching Children about Parts of Speech (Language Book 3)* by Habakkuk Educational Materials can be purchased separately to help students quickly learn the parts of speech. After students have completed this worksheet and turned it in, complete a copy of it on your classroom screen.

Name: ______________________________

Addition and Subtraction Facts

Directions

1. Count the alarm clocks above the first space and write the addend.
2. Count the alarm clocks above the second space and write the addend.
3. Count all the alarm clocks and write the sum after the equal sign.

________ + ________ = ________

Directions

1. Count the alarm clocks above the first space and write the number.
2. Cross out 2 alarm clocks.
3. Count the number of alarm clocks left and write the answer after the equal sign.

________ − 2 = ________

QUESTIONS FOR THE THEMATIC UNITS GAME

Please refer to pages 28-29 of this book for instructions on how to use the unit questions.

Why do we have day and night? *Lesson 15 (Science Themes)*	How long does it take for the earth to spin completely around? *Lesson 15 (Science Themes)*

LESSON 15 TEST (Day and Night)

1. Why do we have day and night?

   ◯ The earth rotates on its axis.

   ◯ When the part of the earth you live on is facing the sun, you have daytime, but it is dark at the same time in another part of the world.

   ◯ Both of the above answers are true.

2. How long does it take for the earth to spin completely around?

   ◯ 12 hours

   ◯ 24 hours

   ◯ 48 hours

Note: These cards can be used for the spelling word games listed on pages 21-22.

meals	moon
muffins	morning
midnight	dream

Note: These cards can be used for the spelling word games listed on pages 21-22.

alarm	pajamas
home	time
daytime	nighttime

Lesson 16

Theme: Holiday Customs (Hanukkah, Christmas, New Year's Eve)
Sound: diphthong /oi/
Spelling Patterns: *oi*, *oy*

Spelling Words	Alphabetized
1. oil	________
2. noise	________
3. noisy	________
4. rejoice	________
5. join	________
6. foil	________
7. coins	________
8. enjoy	________
9. boys	________
10. toys	________
11. joy	________
12. joyful	________

Inflectional endings: enjoys, enjoying, enjoyed

Bonus compound word: tinfoil

Extra challenge word: poinsettia

Thematic Activities for Lesson 16 (Holiday Customs: Hanukkah, Christmas, and New Year's Eve)

Thematic objectives: To familiarize students with the history behind the Jewish Hanukkah celebration and with customs associated with it; to familiarize students with Christmas and New Year's Eve traditions

Key Questions for Discussion

- Hanukkah, Christmas, and New Year's Eve are holidays people celebrate during what month? (*December*)
- What are some Christmas traditions? (*decorating Christmas trees, giving gifts...*)
- What are some Hanukkah traditions? (*lighting one candle of the Menorah the first day of Hanukkah, 2 the second day, and so on until all eight candles are lit by the eighth day of the celebration; eating latkes; playing dreidel; giving gifts; giving chocolate coins to children called gelt*)
- What are some New Year's Eve traditions? (*shooting firecrackers and making lots of other noise at midnight to welcome in the new year*)
- What holiday is celebrated by Jewish people in remembrance of a miracle that took place in the second Temple of Jerusalem after a Jewish family, known as the Maccabees, fought back when their enemies tried to stop them from worshiping God? (*Hanukkah*)
- Why does the Jewish holiday Hanukkah last 8 days? (*God miraculously caused a one day's supply of olive oil to last for 8 days so the lamp in the Temple could stay lit just long enough for them to get some more oil.*)
- Hanukkah (or the "Festival of Lights") is an 8-day Jewish holiday that involves lighting a new candle of the Menorah each day in remembrance of a miracle that happened over 2,000 years ago. What was the miracle? (*See the "What is Hanukkah?" page for details.*)

Activities, Supplies, and Preparations for this Lesson

- **Spelling Patterns of Sounds Heard in the English Language:** Display words with this week's sound on the wall display. (See pages 26-27 for details.)
- **Reproducible materials:** Prepare the reproducible materials provided with this lesson, including the "Spelling Word Riddles," the question cards, and the spelling word cards
  (Supplies: Computer paper, light blue and yellow cardstock, magnetic tape)
- **Books and videos:** Read books and/or watch YouTube videos about Christmas, Hanukkah, and New Year's Eve and about traditions associated with them. (For any books you read, identify the title, the author, and the illustrator. Also, examine the front and back cover of the book before making any predictions about the story or discussing what it is about.)

- **Dreidel:** Play the traditional Hanukkah game dreidel with students. (The toy can be purchased cheaply online at TraditionsJewishGifts.com. Choose one that identifies the names of the letters—*nun*, *gimel*, *hay*, and *shin*.)

  HOW TO PLAY DREIDEL \drā-d^{ə}l\

  There are four Hebrew letters on the dreidel:
  NUN \nun\: do nothing (The letter *u* in *nun* makes the *u* sound in *put*.)
  GIMEL \gĭ-məl\: Take all the money.
  HAY \hā\: Take half the money.
  SHIN \shēn\: Put 2 coins into the middle of the circle.

  Children sit in a circle (you need at least two players). Give each player about ten chocolate flavored coins wrapped in gold foil (these are called *gelt*). Note that you can use play coins in place of chocolate. The game begins by having each player put one coin into the center of the circle. The youngest player is then the first to spin the dreidel. If it lands on, for example, "GIMEL," the spinner takes all the money from the middle of the circle. If it lands on "SHIN," he puts 2 more coins into the middle of the circle, and so forth. Go around the circle, allowing each child a chance to spin the dreidel. Before each spin, however, every child must again put one coin into the center of the circle. Players continue playing until they are out of coins.

  (Supplies: A toy dreidel can be purchased cheaply online at TraditionsJewishGifts.com. Choose one that identifies the names of the letters: *nun*, *gimel*, *hay*, and *shin*. You will also need about ten chocolate flavored coins wrapped in gold foil for each student. These are called *gelt*. You can use play coins in place of chocolate if you choose.)

- **Writing/Handwriting assignment**
  Younger Students: Make a copy of the reproducible writing paper provided on page 30 and position it under a Smart Document Camera or a similar device. Discuss the correct stroke for each letter as you copy the incomplete sentences below (or one of your own choosing) onto your paper. Students should follow along with you, copying the letters on papers of their own. (This will help them to remain on task and to finish the handwriting assignment in unison.) They can then finish the sentence, add additional sentences, and include appropriate illustrations.
  My favorite Christmas tradition is ______.
  My favorite Hanukkah tradition is ______.
  My favorite New Year's Eve tradition is ______.

  Older Students: Older students can write about the topic below. Let them copy the question and respond to it in their writing journals. They may also want to add appropriate illustrations.
  What is your favorite Christmas or Hanukkah tradition and why?

What is Hanukkah?

Hanukkah (also known as the "Festival of Lights") is a Jewish Holiday celebrated around Christmastime. It lasts eight whole days! The holiday reminds the Jewish people of a time over 2,000 years ago when their enemies abused their Holy Temple and tried to stop them from worshiping God. However, a brave Jewish family, known as the Maccabees, led a revolt against the enemy and stopped their misuse of the Temple. Hanukkah is celebrated in remembrance of their rededication of the second Temple of Jerusalem and of the miracle that took place there.

When the Jewish people took back the Holy Temple, the place where they worshiped God, they knew they needed to relight the Menorah \mə-nor-ə\ because God had already commanded His people to keep the lamp burning. Although they found enough olive oil to relight the lamp, there was only enough oil to keep it burning in the Temple for one day. Miraculously, however, God didn't let the oil run out. It ended up lasting eight whole days, which gave them exactly enough time to get more oil. God was showing His people that He was with them.

In remembrance of this miracle, one candle is lit the first day of Hanukkah, two the second day, and so forth. By the eighth day of the celebration, all eight candles are lit. A Hanukkah menorah holds nine candles, but the middle candle, called the Shamash \shŏməs\, is just used to light the others. Besides lighting the candles, boys and girls are often given gifts and treated with chocolate coins wrapped in gold foil during the celebration. The chocolate coins are called gelt. It is also tradition to eat latkes \lŏt-kəz\ and to play a game called dreidel during Hanukkah.

Context Clues

Directions: Fill in each blank with a spelling word that makes sense in the context of the sentence. (Cross out the words from the list as you use them.) Circle the letters in the key words that make the /oi/ sound (*oi*, *oy*).

Spelling words: oil, noise, noisy, rejoice, join, foil, coins, enjoy, boys, toys, joy, joyful

1. Christmas, Hanukkah, and New Year's Eve are holidays people ____________ during December.

2. Shopping centers are crowded and ____________ this time of year.

3. Parents are looking for ____________ for their kids.

4. Everyone seems to have so much ____________.

5. Children are especially ____________ this time of year.

6. **Hanukkah** is an eight-day Jewish holiday that involves lighting candles in remembrance of a miracle about ____________. (Read the next page to learn more!)

7. ____________ and girls may get to help light the candles.

8. Besides lighting candles, the children also get chocolate ____________ called *gelt*.

9. The chocolate coins are wrapped in gold ____________.

10. About a week after Christmas on **New Year's Eve**, family and friends ____________ together to celebrate.

11. We shoot firecrackers and make lots of ____________ at midnight.

12. We ____________ because it is a new year.

Spelling Word Riddles

Discussion: All spelling words this week have the /oi/ sound and can be spelled as ***oi*** as in *oil* or ***oy*** as in *boy*. (Help students hear and pronounce the /oi/ sound by first saying the sound and then a basic word, such as *oil*.)

Directions: Reproduce the next page on light blue and yellow cardstock. Cut out the pieces and attach a magnetic strip to the back of each piece. Line up the yellow pieces horizontally towards the top of a magnetic board. Position the light blue strips vertically below them. (See the illustration on page 15.) Read a riddle and call on a student to identify the correct spelling word. For the first sentence, a student would identify the word *foil* and then move the yellow magnetic strip with the letters ***oi*** between the light blue letters ***f*** and ***l*** to form the word ***foil***.

1. In this word, the /oi/ sound is spelled *o-i*. Chocolate coins called gelt and other kinds of chocolate are wrapped in this. (*Foil*)
2. In this word, the /oi/ sound is spelled *o-i*. It means the opposite of quiet. (*Noisy*)
3. In this word, the /oi/ sound is spelled *o-y*. It is the antonym of girls. (*Boys*)
4. In this word, the /oi/ sound is spelled *o-i*. Hanukkah is an eight-day Jewish holiday that involves lighting candles in remembrance of a miracle about this. (*Oil*)
5. In this word, the /oi/ sound is spelled *o-y*. Children like to play with these. (*Toys*)
6. In this word, the /oi/ sound is spelled *o-i*. It begins with *j* and rhymes with coin. (*Join*)
7. In this word, the /oi/ sound is spelled *o-y*. It means "full of joy." (*Joyful*)
8. In this word, the /oi/ sound is spelled *o-i*. These can be used to buy things with. (*Coins*)
9. In this word, the /oi/ sound is spelled *o-y*. It rhymes with *boy* and means "happiness." (*Joy*)
10. In this word, the /oi/ sound is spelled *o-i*. This is something you can hear with your ears. (*Noise*)
11. In this word, the /oi/ sound is spelled *o-y*. The prefix *en-* is attached to this word and can mean that you like doing something. (*Enjoy*)
12. In this word, the /oi/ sound is spelled *o-i*. This is what you do when you open a present you have been wanting. (*Rejoice*)

l	n	se	n	sy
rej	ce	j	n	j
f	l	c	ns	
enj	b	s		
t	s	j	ful	

Directions: Reproduce the cards above on light blue cardstock and the cards below on yellow cardstock. Cut out the pieces and attach magnetic strips to the backs. (See the previous page for further instructions.)

oy	oy	oy	oy	oy

oi	oi	oi	oi	oi	oi	oi

Spelling Patterns of Sounds Heard in Our Spelling Words

Directions: Fill in the missing letters that make the /oi/ sound.

- Use a **blue** colored pencil for ***oi***.
- Use a **red** colored pencil for ***oy***.

_ _ l	j _ _ n	b _ _ s
n _ _ s e	f _ _ l	t _ _ s
n _ _ s y	c _ _ n s	j _ _
r e j _ _ c e	e n j _ _	j _ _ f u l

Directions: Record spelling words whose /oi/ sound is spelled *oi* under the **"oi"** headings and words whose /oi/ sound is spelled *oy* under the **"oy"** heading.

oi	oi	oy

Spelling Words (/oi/ sound)

R	E	J	O	I	C	E	W	E	J	U	C
U	S	L	I	L	K	S	J	T	O	B	O
G	U	O	D	M	R	Y	O	E	I	J	I
C	T	B	M	H	T	E	Y	D	N	H	N
N	O	I	S	Y	N	R	D	J	Z	D	S
B	I	N	F	J	O	O	A	L	M	N	L
J	D	E	Q	V	I	M	T	O	Y	S	J
O	F	N	Z	Q	L	H	P	K	G	T	E
Y	S	J	W	E	G	H	B	C	R	F	O
F	G	O	Y	B	O	Y	S	H	I	O	C
U	R	Y	E	I	P	X	Y	V	F	I	R
L	Y	L	N	O	I	S	E	J	P	L	X

Holiday Customs

1. oil
2. noise
3. noisy
4. rejoice
5. join
6. foil
7. coins
8. enjoy
9. boys
10. toys
11. joy
12. joyful

Directions: Add the prefix and suffixes listed below to the root word *joy* to form a word that means "to like something," "full of joy," and "not having any joy."

Prefix: *en-*
Suffixes: *-ful*, *-less*

Prefix	Root Word	Suffix
__________	joy	__________
__________	joy	__________
__________	joy	__________

Which word means "to like something"? ____________________

Which word means "full of joy"? ____________________

Which word means "not having any joy"? ____________________

Directions: Insert a prefix or suffix in the space provided to form a new word that makes sense in the context of the sentence.

Do you feel **joy** around Christmastime? Most people do _____**joy** the holidays. Children are especially **joy**_____ this time of the year. For some people, though, it is a sad time, and they feel **joy**_____.

- A **prefix** is added to the beginning of a root word to form a new word (altogether). A **suffix** is added to the end of a root word to form a new word (thankful). Both prefixes and suffixes are **affixes**.
- **Derivative:** A root word with something added to it (e.g., a prefix or a suffix)
- **Rule:** When a consonant suffix is added to a word, just put the word and the suffix together. (Thankful)

Weekly Oral Language

1. Correct the asking sentence.

   how many days does hanukkah last

2. Correct the telling sentence.

   hanukkah is celebrated for eight days

3. What is the subject of the telling sentence?

   ____

4. Read the *telling sentence* again and identify the helping verb and the verb (action word).

   Helping Verb: ______________________

   Verb: ______________________________

5. Write your own sentence about this week's theme.

   ____

Instruction: *Teaching Children about Capitalization, Punctuation, Apostrophe s, and Contractions (Language Book 2)* by Habakkuk Educational Materials can be purchased separately to give students the knowledge they need to complete this worksheet. After students have completed the worksheet and turned it in, complete a copy of it on your classroom screen.

Parts of Speech

Directions: Identify what parts of speech are used in the sentence below and write them in the spaces provided. Then identify the subject (naming part) and predicate (action part) of the sentence and write them in the spaces provided.

Parts of speech: Noun marker, noun, pronoun, verb, helping verb, adjective, adverb, preposition, conjunction, interjection

Candles	are	lit	during	Hanukkah.
___	___	___	___	___

Subject (Naming Part)

____

Predicate (Action Part)

____

Instruction: *Teaching Children about Parts of Speech (Language Book 3)* by Habakkuk Educational Materials can be purchased separately to help students quickly learn the parts of speech. After students have completed this worksheet and turned it in, complete a copy of it on your classroom screen.

Diphthongs *ou, ow, oi,* and *oy*

Spelling with *ou* and *ow*: The diphthongs *ou* as in *out* and *ow* as in *cow* make the same sound, but

- ***ou*** is usually found at the beginning of a word or syllable or in the middle of it (out, shout), whereas
- ***ow*** is usually found at the end of a word (cow).

Spelling with *oi* and *oy*: The diphthongs *oi* as in *oil* and *oy* as in *boy* make the same sound, but

- ***oi*** is usually found at the beginning of a word or syllable or in the middle of it (oil, coin), whereas
- ***oy*** is usually found at the end of a word (boy).

Out, Sh**ou**t, c**ow**	**Oi**l, C**oi**n, B**oy**
ou_____	oi_____
____ou____	____oi____
_____ow	_____oy

____

____

Diphthong: Two vowel sounds that come together so fast that they are considered one syllable

Examples: ***ou*** (ouch, our, found, house); ***ow*** (how, now); ***oi*** (oink, join, point); ***oy*** (joy, toy)

Name: ______________________

Addition and Subtraction Facts

Directions

1. Count the presents above the first space and write the addend.
2. Count the presents above the second space and write the addend.
3. Count all the presents and write the sum after the equal sign.

______ + ______ = ______

Directions

1. Count the Christmas ornaments above the first space and write the number.
2. Cross out 7 Christmas ornaments.
3. Count the number of ornaments left and write the answer after the equal sign.

______ − 7 = ______

Addition and Subtraction Word Problems

Directions: Color and cut out the pictures at the bottom of the page. Read each sentence and use the cutouts to help solve the problem. Write the addition or subtraction fact in the spaces provided.

1. I had 2 Christmas presents. My uncle gave me 1 more. How many presents do I have in all?

   ___________ = _____

2. On Christmas morning, there were 10 wrapped presents under the tree. Six presents have already been opened. How many presents are still wrapped?

   ___________ = _____

3. There were 8 ornaments on the Christmas tree. Six fell off. How many ornaments are left on the tree?

   ___________ = _____

QUESTIONS FOR THE THEMATIC UNITS GAME

Please refer to pages 28-29 of this book for instructions on how to use the unit questions.

Hanukkah, Christmas, and New Year's Eve are holidays people celebrate during what month? *Lesson 16 (Social Studies)*	What are some Christmas traditions? *Lesson 16 (Social Studies)*
What are some Hanukkah traditions? *Lesson 16 (Social Studies)*	What are some New Year's Eve traditions? *Lesson 16 (Social Studies)*
What holiday is celebrated by Jewish people in remembrance of a miracle that took place in the second Temple of Jerusalem after a Jewish family, known as the Maccabees, fought back when their enemies tried to stop them from worshiping God?	Why does the Jewish holiday Hanukkah last 8 days? *Lesson 16 (Social Studies)*
Hanukkah (or the "Festival of Lights") is an 8-day Jewish holiday that involves lighting a new candle of the Menorah each day in remembrance of a miracle that happened over 2,000 years ago. What was the miracle? *Lesson 16 (Social Studies)*	

LESSON 16 TEST (Holiday Customs: Hanukkah, Christmas, New Year's Eve)

1. What day is Christmas celebrated on?
   - ○ July 4
   - ○ October 31
   - ○ December 25

2. Why is Christmas celebrated?
   - ○ To remember the birth of Jesus
   - ○ To remember a miracle about oil
   - ○ To celebrate the winter season

3. Hanukkah, Christmas, and New Year's Eve are holidays people celebrate during what month?
   - ○ October
   - ○ November
   - ○ December

4. What are Christmas traditions? (Mark TWO answers.)
   - ○ Decorating Christmas trees
   - ○ Giving gifts
   - ○ Playing dreidel

5. What are Hanukkah traditions? (Mark TWO answers.)
   - ○ Having your picture taken with Santa
   - ○ Lighting a new candle of the Menorah daily for 8 days
   - ○ Playing dreidel

6. What is a New Year's Eve tradition?

- ◯ Playing dreidel
- ◯ Shooting firecrackers and shouting "Happy New Year" at midnight
- ◯ Hunting for eggs

7. What holiday is celebrated by Jewish people in remembrance of a miracle that took place in the second Temple of Jerusalem after a brave Jewish family, known as the Maccabees, fought back when their enemies tried to stop them from worshiping God?

- ◯ Hanukkah
- ◯ Christmas
- ◯ New Year's Eve

8. Hanukkah (or the "Festival of Lights") is an 8-day Jewish holiday that involves lighting a new candle of the Menorah each day in remembrance of a miracle that happened over 2,000 years ago. What was the miracle?

- ◯ God spread apart the waters of the Red Sea.
- ◯ God caused water to come out of a rock when His people were thirsty.
- ◯ God miraculously caused a one day's supply of oil to last for 8 days so that the lamp in the Temple could stay lit as God commanded.

Note: These cards can be used for the spelling word games listed on pages 21-22.

o̲i̲l	no̲i̲se
no̲i̲sy	rejo̲i̲ce
jo̲i̲n	fo̲i̲l

Note: These cards can be used for the spelling word games listed on pages 21-22.

coins	enjoy
boys	toys
joy	joyful

Lesson 17

Theme: The Birth of Christ (Biblical theme)
Sound: Long *e* /ē/
Spelling Patterns: *e*, *ea*, *ee*, *y*, *ey*

Spelling Words	Alphabetized
1. Jesus	______________
2. east	______________
3. peace	______________
4. meet	______________
5. see	______________
6. sheep	______________
7. baby	______________
8. city	______________
9. family	______________
10. happy	______________
11. Mary	______________
12. donkey	______________

Extra challenge words: country, nativity

Note: The initial letters of *Mary* and *Jesus* must be capitalized.

Thematic Activities for Lesson 17
(The Birth of Christ)

Thematic Objectives: Students will understand the true meaning of Christmas. They will learn of the birth of Jesus, events that determined the place of His birth, and events that occurred following His birth. They will also understand the meaning of the words *census*, *stable*, *swaddling clothes*, and *manger*.

Key Questions for Discussion

- What day is Christmas celebrated on? (*December 25*)
- Why is Christmas celebrated? (*to remember the birth of Jesus*)
- Who was Jesus' mother? (*Mary*)
- Where was Jesus born? (*in a stable in the city of Bethlehem*)
- Why was Jesus born in Bethlehem even though his mother and earthly father lived in a different city? (*Mary and Joseph had to travel there to register for a census. They went there to be counted.*)
- Why was Jesus born in a stable where animals were kept? (*There was no room for them in the inn.*)
- How did the shepherds who were watching their sheep learn about the birth of Jesus and where to find Him? (*from angels*)
- What did the shepherds do after they met the newborn Savior? (*They spread the news about His birth.*)

Activities, Supplies, and Preparations for this Lesson

- **Spelling Patterns of Sounds Heard in the English Language:** Display words with this week's sound on the wall display. (See pages 26-27 for details.)
- **Reproducible materials:** Prepare the reproducible materials provided with this lesson, including the "Spelling Word Riddles," the question cards, and the spelling word cards
  (<u>Supplies:</u> Computer paper, light blue and yellow cardstock, magnetic tape)
- **Books and videos:** Read books and/or watch YouTube videos that discuss the true meaning of Christmas, such as "A Charlie Brown Christmas." (For any books you read, identify the title, the author, and the illustrator. Also, examine the front and back cover of the book before making any predictions about the story or discussing what it is about.)
- **Crafts:** Choose Christmas craft activities for kids from online and decorate the room with their creations.

- **Writing assignment (Christmas cards) and a field trip to a local nursing home or children's hospital:** Reproduce the Christmas cards included with this lesson on white cardstock. Instruct students to choose one or more cards to color and cut out. Let them suggest possible phrases to write on the front of the card, such as "Merry Christmas!" or "Merry Christmas and Happy New Year!" Record their suggestions on the whiteboard. Then let them write notes to the recipients of the cards. Possibly, the children could make cards to be delivered to local nursing home residents or to patients in a children's hospital. They might tell the patients who they are, what school they go to, and their grade or age. Let students suggest other statements to include inside the cards, such as "God Loves You!" or "Have a blessed holiday!" If possible, plan a field trip so that students can deliver the cards themselves and sing a few Christmas carols to the patients.
(Supplies: White cardstock; signed permission slips for each child to attend a field trip to a local nursing home or children's hospital to deliver Christmas cards and possibly sing a few Christmas carols)
- **Geography:** Show your class a world map. Ask students to point out the country or state where they live and mark the location with a thumbtack or by circling it. Then mark the city of Bethlehem while explaining to students that this was the place where Jesus was born. (Continue to the vocabulary activity below.)
- **Vocabulary (census, swaddling clothes, stable, manger):** Print the words *census*, *swaddling clothes*, *stable*, and *manger* on the whiteboard.

Census: Help students to understand that Mary and Joseph left their home in Nazareth to travel to Bethlehem because they had to register for a census ordered by Caesar Augustus. A census involves counting all the people who live in a country, and Caesar Augustus wanted everyone who lived in the Roman Empire to be counted. You might want to conduct your own "census" by counting the students in your classroom or in all the classrooms of your school.

Swaddling clothes: Explain to students that Jesus' mother, Mary, wrapped her baby in strips of cloth called swaddling clothes. Images of swaddling clothes are available online to show your students.

Stable: Explain to students that Jesus was born in a stable because there was no room for them in the inn. A stable was like a barn where the animals were kept.

Manger: Tell students that Mary laid Baby Jesus in a manger. Show them a picture from online while explaining that a manger was a box that the animals ate from.

- **"The Birth of Christ" story:** Have students read "The Birth of Christ" story included with this lesson. Read over the questions that follow it first so that students will know what to look for while reading. If you follow the link beside "Story Illustrations" under the title, you will see several slides about the story that can be displayed as students read. After reading, let the students answer the questions individually or as a class. These same questions can be used for the "Christmas Story Game" described below.
- **Christmas Story Game:** Ask students to sit in a circle on the gathering rug. Give one child a Christmas bag filled with different flavors of candy canes (or whatever else you choose). Play Christmas music and instruct students to pass the bag clockwise to the child next to him/her. After several seconds, pause the music and ask the child holding the bag one of the Christmas story questions from "The Birth of Christ" story included with this lesson (and/or a question listed under "Key Questions for Discussion" from page 309). If the child answers the question correctly, let him or her have one of the candy canes from inside. Continue until everyone has a cane. (As a variation to filling the bag with candy canes, you might just use it to hold the story questions from page 334 and let students who are holding the bag when the music stops draw one from the bag to read.)
  (Supplies: A Christmas gift bag filled with different flavors of candy canes or whatever other treat you choose; Christmas music)
- **Christmas Party (or a Birthday Party for Jesus):** During the party, you might want to let students watch a Christmas cartoon from YouTube and to open any gifts they may have. (Teachers sometimes wrap small gifts for their students, like Christmas pencils, Play-Doh, or inexpensive books.) There are also plenty of Christmas party game ideas available online if you want to incorporate some into your party. Possible snacks to provide for the party are listed on the note for parents on the following page.

  (Supplies: Send parents a note about the Christmas party as soon as possible so they can help provide the snacks and party supplies. A note with information about the party is provided on the following page. Gather supplies for any Christmas games from online you choose to play during your celebration.)

Dear Families,

Here is what you need to know about our classroom Christmas party. Please return this note to me, indicating if you would like to volunteer and/or donate something for the party by signing your name beside the item. If what you want to bring is not listed, please record it in one of the blank boxes.

Number of students:

Date:

Time:

	Parent's Signature
I am available to assist during the party.	
Paper plates and napkins	
Store-bought Christmas cupcakes	
Store-bought Christmas cookies	
Capri Sun or other packaged juice	

The Birth of Christ

(Micah 5:2; Luke 2:1-7)

Caesar Augustus	Micah	Nazareth
\sē-zər ə-gəs-təs\	\mī-kə\	\nă-zə-rəth\

Hundreds of years before Jesus was born, God promised through the prophet Micah that the Savior, who is Jesus, would be born in Bethlehem. However, Mary, who was pregnant with Jesus, and Joseph, the man she was going to marry, did not live in Bethlehem. They lived in the town of Nazareth, so how could Jesus be born in Bethlehem? God used the Roman Emperor Caesar Augustus to help get them there! (Note: Several decades before Jesus was born, Rome conquered Israel, and it became a part of the Roman Empire and stayed that way for several hundred years. Thus, the events that occurred during the New Testament took place while Israel was controlled by the Romans rather than the Jews like it is today.)

Caesar Augustus wanted everyone who lived in the Roman Empire to be counted, so he ordered them to return to the cities where their families were from to register for a census. Since Joseph's family was from Bethlehem, he and Mary had to travel there. Imagine how difficult this journey must have been for Mary when she was carrying a baby in her stomach! She was going to give birth to God's Son!

When they finally reached Bethlehem, they needed a place to stay. However, they couldn't stay in the inn because there was no room for them. Instead, they had to stay in a stable where the animals were kept, and that's where Baby Jesus was born. Mary wrapped her Baby in strips of cloth called swaddling clothes and laid Him in a box that the animals ate from. The box was called a manger. Just as God had told the prophet Micah hundreds of years before, the Savior was born in Bethlehem, and we celebrate His birth every year at Christmas.

The day Jesus was born, an angel visited some shepherds. Shepherds take care of sheep. While they were watching their sheep at night, the angel appeared. He was surrounded by a bright light. The shepherds were afraid, but the angel said: "Do not be afraid! I have good news! Today the Savior, which is Christ the Lord, was born in the City of David!" (The City of David is another name for Bethlehem or Jerusalem, but Jesus was born in Bethlehem.) The angel even told the Shepherds how they would be able to recognize the Savior. He said, "The Baby will be wrapped in swaddling clothes and lying in a manger." Then more angels came, and together the angels sang praise songs to God.

After the angels went back to heaven, the shepherds hurried to Bethlehem to find the Savior. When they found Joseph and Mary at the stable with Baby Jesus who was lying in a manger, they worshipped Him. Later, as they returned to their flock, they thanked God for everything they had seen and heard, and they told everyone about the new Savior.

Note: Stories that tell about Jesus are found in the New Testament, and this story is from the Gospel of Luke.

1. Where in the Bible is this story found? ______________________________
2. Who is the Savior? ______________________________
3. What Old Testament prophet prophesied that the Savior would be born in Bethlehem?

   ____
4. Why did Joseph and Mary have to leave their home in Nazareth to travel to Bethlehem where Joseph's family was from? ______________________________

   ____
5. Why did Caesar Augustus order the census? ______________________________

   ____
6. What empire controlled Israel during New Testament times? ______________
7. Where in Bethlehem did Mary and Joseph find a place to stay? ______________

8. What is a stable? ______________________________

9. Who was born in the stable? ______________________________

10. What did Mary wrap Baby Jesus in? ______________________________

11. A manger was used as a bed for Baby Jesus. What was a manger normally used as?

____

12. Why do we celebrate Christmas? ______________________________

13. What do shepherds do? ______________________________

14. Who came to the shepherds one night with good news? ______________________________

15. What good news was told to the shepherds? ______________________________

____

16. When more angels came to the shepherds, what did the angels do? ______________________________

____

17. Where did the shepherds find the Savior? ______________________________

____

18. How did the shepherds know where to find Baby Jesus? ______________________________

19. Bethlehem and Jerusalem are both known as "The City of David." Which of these cities was Jesus born in? ______________________________

20. What did the shepherds do when they found Baby Jesus? ______________________________

____

21. What two things did the Shepherds do after they left Baby Jesus? ______________________________

____

Context Clues

Directions: Fill in each blank with a spelling word that makes sense in the context of the sentence. (Cross out the words from the list as you use them.) Circle the letters in the key words that make the long *e* sound (*e*, *ea*, *ee*, *y*, *ey*).

Spelling words: Jesus, east, peace, meet, see, sheep, baby, city, family, happy, Mary, donkey

1. Christmas is celebrated every December 25th to remember the birth of the __________ Jesus.

2. Jesus was born in the __________ of Bethlehem.

3. His mother, __________, and earthly father Joseph lived in another city but had to travel there to register for a census.

4. Mary may have ridden on a __________ to get there.

5. In Bethlehem, there was no room for them in the inn, so they had to stay in a stable with the __________ and other animals.

6. That's where Baby __________ was born.

7. In a different country, some Wise Men saw a star shining in the __________.

8. They were __________ to see the star because they understood it meant that the King of the Jews had been born.

9. They followed the star all the way to Bethlehem and met Jesus and His __________.

10. An angel also appeared in the sky and told some shepherds where to go to __________ Baby Jesus.

11. Then more angels came and sang, "Glory to God in the highest, and on earth __________, good will toward men" (Luke 2:14, KJV).

12. The shepherds were happy to __________ the newborn Savior and helped to spread the news about His birth.

Spelling Word Riddles

Discussion: All spelling words this week have the long *e* sound, /ē/, and can be spelled as ***e*** as in *Jesus*, ***ea*** as in *east*, ***ee*** as in *sheep*, ***y*** as in *baby*, or ***ey*** as in *donkey*. The long *e* sound is the same as the name of the letter *e*. In the word *east*, what vowel sound do you hear? (Help students hear and pronounce the long *e* sound by first saying "*e*" and then a basic word, such as *east*.)

Directions: Reproduce the next two pages on light blue or yellow cardstock as instructed. Cut out the pieces and attach a magnetic strip to the back of each piece. Line up the yellow pieces horizontally towards the top of a magnetic board. Position the light blue strips vertically below them. (See the illustration on page 15.) Read a riddle and call on a student to identify the correct spelling word. In the example riddle, a student would identify the word *tree* and then move the yellow magnetic strip with the letters ***ee*** to the right of the light blue magnetic strip having the initial letters ***tr*** to form the word ***tree***.

Example: In this word, the long *e* sound is spelled *e-e*. Ian helps to decorate this at Christmas time. (Move the red *ee* strip next to the green *tr* strip to spell the word *tree*.)

1. In this word, the long *e* sound is spelled *y*. The angel Gabriel told this person that she would be the mother of God's Son. (*Mary*)
2. In this word, the long *e* sound is spelled *y*. Mary felt this way after the angel Gabriel told her that she would be the mother of God's Son. (*Happy*)
3. In this word, the long *e* sound is spelled *e-y*. Mary may have ridden upon this animal to get to Bethlehem. (*Donkey*)
4. In this word, the long *e* sound is spelled *e*. He is God's Son. (*Jesus*)
5. In this word, the long *e* sound is spelled *y*. Jesus was born in the _____ of Bethlehem. (*City*)
6. In this word, the long *e* sound is spelled *e-e*. Shepherds were watching these the night Jesus was born. (*Sheep*)
7. In this word, the long *e* sound is spelled *e-a*. The shepherds heard the angels sing, "Glory to God in the highest, and on earth _____, good will toward men" (Luke 2:14, KJV). (*peace*)
8. In this word, the long *e* sound is spelled *e-a*. The Wise Men saw a new star shining in this direction. (*East*)
9. In this word, the long *e* sound is spelled *y*. Jesus was not a man when the Wise Men brought Him gifts; He was a _____. (*baby*)
10. In this word, the long *e* sound is spelled *y*. Joseph, Mary, and Jesus were a _____. (*family*)
11. In this word, the long *e* sound is spelled *e-e*. It begins like *Mary* and rhymes with *eat*. (*Meet*)
12. In this word, the long *e* sound is spelled *e-e*. Michael does this with his eyes. (*See*)

Directions: Use light blue cardstock to reproduce this page. Cut out the pieces and attach magnetic strips to the back of each piece. (See the previous page for further instructions.)

sus	st	tr	J
p	ce	m	t
s	sh	p	
bab	cit		
famil	happ		
Mar	donk		

Directions: Use yellow cardstock to reproduce this page. Cut out the pieces and attach magnetic strips to the backs.

ea	ea	ea	ea	ey	e
ee	ee	ee	ee	ey	e
y	y	y	y	y	e

Spelling Patterns of Sounds Heard in Our Spelling Words

Directions: Fill in the missing letters that make the long *e* sound /ē/.

- Use a **blue** colored pencil for ***e***.
- Use a **red** colored pencil for ***ea***.
- Use a **green** colored pencil for ***ee***.
- Use a **purple** colored pencil for ***y***.
- Use a **brown** colored pencil for ***ey***.

J_sus	s__	famil_
__st	sh__p	happ_
p__ce	bab_	Mar_
m__t	cit_	donk__

Directions: Record spelling words whose long *e* sound is spelled *e* beside the **"e"** heading, words whose /ē/ sound is spelled *ea* beside the **"ea"** heading, and so forth.

e ______________________________

ea ______________________________

ee ______________________________

y ______________________________

ey ______________________________

Spelling Words (Long *e* /ē/ sound)

K	C	N	V	B	N	A	W	E	G	D	O
U	S	L	F	A	M	I	L	Y	X	O	F
S	E	E	T	D	A	R	O	E	M	N	V
C	T	B	H	A	P	P	Y	D	X	K	Z
J	P	K	C	H	N	R	D	M	Z	E	A
E	I	N	E	A	S	T	A	E	M	Y	L
S	A	D	Q	V	K	M	W	E	S	Z	J
U	F	N	Z	Q	F	H	P	T	G	T	S
S	S	B	A	B	Y	H	B	C	R	W	H
X	G	Y	Y	W	M	M	A	R	Y	Q	E
I	R	Q	E	I	P	X	Y	V	F	M	E
P	E	A	C	E	Z	C	I	T	Y	J	P

The Birth of Christ

1. Jesus
2. east
3. peace
4. meet
5. see
6. sheep
7. baby
8. city
9. family
10. happy
11. Mary
12. donkey

Homophones

Mary	marry	merry

Mary is the mother of Jesus.	The couple will **marry** in a church.	**Merry** Christmas!
Mary is a girl or a woman's name.	*To marry is to become husband and wife.*	*Merry means "happy."*

Mary felt **merry** when Joseph asked her to **marry** him.

Underline the word that refers to a girl or to a woman's name.
Circle the word that means "to become husband and wife."
Highlight the word that means "happy."

Homophones

meat	meet

Turkey is a favorite holiday **meat**.

Meat is a food.

It's so nice to **meet** you!

Meet is coming to know a person or getting together with them.

Let's **meet** at the **meat** shop to buy some steak for dinner.

Underline the word that refers to a food.
Circle the word that refers to coming to know a person or getting together with them.

Homophones

piece

I got the last **piece** of pizza.

A piece is a part of something.

peace

A dove is often used
as a symbol of **peace**.

Peace means to be free from conflict or to feel good about something.

I don't feel **peace** about
eating the last **piece** of pizza.

Underline the word that means a part of something.
Circle the word that refers to feeling good about something.

Homophones

see

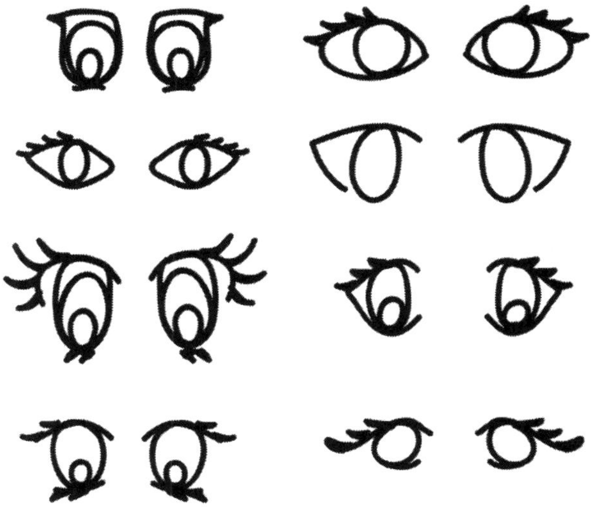

God gave you eyes to **see** with.

See is something that you do with your eyes.

sea

The **sea** is full of ocean animals.

A sea is part of the ocean.

Do you **see** the jellyfish in the **sea**?

Underline the word that refers to what you do with your eyes.
Circle the word that refers to a part of the ocean.

Weekly Oral Language

1. Correct the asking sentence.

   what is christmas

2. Correct the telling sentence.

   christmas is a celebration of jesus' birth

3. What is the subject of the telling sentence?

   ____

4. Read the *telling sentence* again and identify the verb (action word).

   ____

5. Write your own sentence about this week's theme.

   ____

Instruction: *Teaching Children about Capitalization, Punctuation, Apostrophe s, and Contractions (Language Book 2)* by Habakkuk Educational Materials can be purchased separately to give students the knowledge they need to complete this worksheet. After students have completed the worksheet and turned it in, complete a copy of it on your classroom screen.

Parts of Speech

Directions: Identify what parts of speech are used in the sentence below and write them in the spaces provided. Then identify the subject (naming part) and predicate (action part) of the sentence and write them in the spaces provided.

Parts of speech: Noun marker, noun, pronoun, verb, helping verb, adjective, adverb, preposition, conjunction, interjection

Jesus	**was**	**born**	**in**	**Bethlehem.**
__________	__________	__________	__________	__________

Subject (Naming Part)

____

Predicate (Action Part)

____

Instruction: *Teaching Children about Parts of Speech (Language Book 3)* by Habakkuk Educational Materials can be purchased separately to help students quickly learn the parts of speech. After students have completed this worksheet and turned it in, complete a copy of it on your classroom screen.

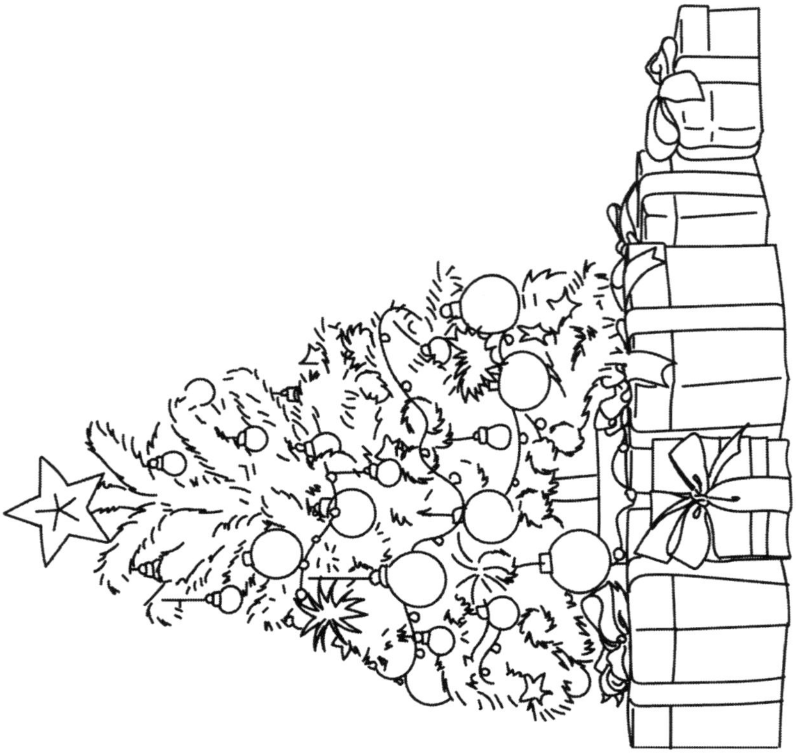

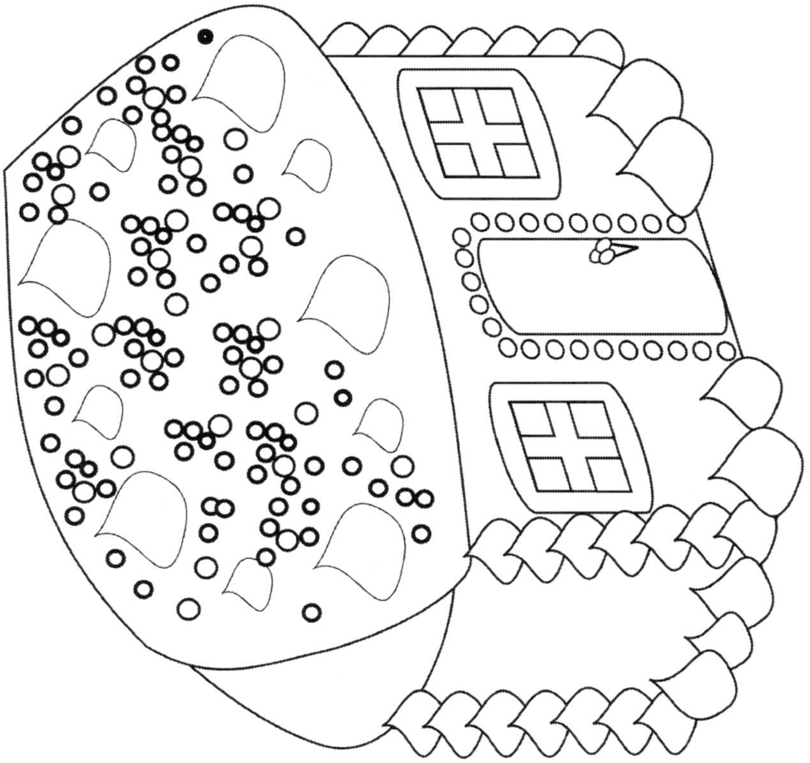

Name: ______________________________

Addition and Subtraction Facts

Directions

1. Count the angels above the first space and write the addend.
2. Count the angels above the second space and write the addend.
3. Count all the angels and write the sum after the equal sign.

______ + ______ = ______

Directions

1. Count the angels above the first space and write the number.
2. Cross out 0 angels.
3. Count the number of angels left and write the answer after the equal sign.

______ − 0 = ______

QUESTIONS FOR THE THEMATIC UNITS GAME

Please refer to pages 28-29 of this book for instructions on how to use the unit questions.

What day is Christmas celebrated on? *Lesson 17 (Social Studies)*	Why is Christmas celebrated? *Lesson 17 (Social Studies)*
Who was Jesus' mother? *Lesson 17 (Social Studies)*	Where was Jesus born? *Lesson 17 (Social Studies)*
Why was Jesus born in Bethlehem even though his mother and earthly father lived in a different city? *Lesson 17 (Social Studies)*	Why was Jesus born in a stable where animals were kept? *Lesson 17 (Social Studies)*
In a different country, some Wise Men saw a star shining in the east. What did the star mean to the Wise Men? *Lesson 17 (Social Studies)*	How did the shepherds who were watching their sheep learn about the birth of Jesus and where to find Him? *Lesson 17 (Social Studies)*
What did the shepherds do after they met the newborn Savior? *Lesson 17 (Social Studies)*	

Note: These cards can be used for the spelling word games listed on pages 21-22.

Jesus	east
peace	meet
see	sheep

Note: These cards can be used for the spelling word games listed on pages 21-22.

bab<u>y</u>	cit<u>y</u>
famil<u>y</u>	happ<u>y</u>
Mar<u>y</u>	donk<u>ey</u>

Lesson 18$_1$

Optional themes for /ō/ words:
The Winter Season (18$_1$),
Noah's Ark (18$_2$)

Theme: The Winter Season
Sound: Long *o* /ō/
Spelling Patterns: *o* with final *e, o, oa, ow*

Spelling Words

1. close
2. froze
3. cold
4. cocoa
5. coat
6. roads
7. below
8. blow
9. slowly
10. snow
11. window
12. throw

Alphabetized

____

____

____

____

____

____

____

____

____

____

____

____

Inflectional endings: snows, snowing, snowed; slowly, slower, slowest

Bonus compound words: overcoat, snowball, snowcap, snowfall, snowflake, snowman, snowstorm, snowsuit, windowsill, polar bear

Thematic Activities for Lesson 18
(The Winter Season)

Thematic objectives: Students will understand that winter is one of the four seasons and will be able to identify animal behavior, weather conditions, and activities that occur in winter. In addition, they will discover that the temperature must be at 32° or below for water to freeze, thus explaining why we often have snow or ice in wintertime rather than rain.

Key Questions for Discussion

- Winter is one of the four seasons. What are the four seasons? (*winter, spring, summer, fall*)
- What happens in nature during the winter season? (*It can get very cold and snow in winter. Many animals hibernate in the winter.*)
- What do animals like bears, bats, and skunks do in winter? (*hibernate*)
- At what temperature does rain change to ice or snow? (*When the temperature is at or below 32°, rain changes to ice or snow.*)
- What are the winter months? (*Part of December, all of January and February, part of March*)

Activities, Supplies, and Preparations for this Lesson

- **Spelling Patterns of Sounds Heard in the English Language:** Display words with this week's sound on the wall display. (See pages 26-27 for details.)
- **Reproducible materials:** Prepare the reproducible materials provided with this lesson, including the "Spelling Word Riddles," the question cards, and the spelling word cards
  (Supplies: Computer paper, light blue and yellow cardstock, magnetic tape)
- **Books and videos:** Read books and/or watch YouTube videos about hibernation, seasons, and winter. Have Fun Teaching has a cute video available on YouTube titled "Seasons Song" that can help students learn about the four seasons. (For any books you read, identify the title, the author, and the illustrator. Also, examine the front and back cover of the book before making any predictions about the story or discussing what it is about.)
- ***The Mitten* book:** Read *The Mitten* to students. There is a version by Alvin Tresselt and Jan Brett. (If you do not have the book, you can hear it read aloud on YouTube.) Tell students that they are going to play a game after the story and will need to remember what animals are mentioned in the book.
  (Supplies: Have available *The Mitten* by Alvin Tresselt or Jan Brett or be prepared to listen to it read aloud on YouTube.)

- **The Mitten Game:** Cut out the mittens on page 352. (You may want to let a student color them first.) Tape them to a lunch sack. Cut out the animal name cards. Students will, in turn, draw a card and read aloud the animal name that's printed on it. If the animal was mentioned in the story, they will put the card in the mitten bag. (**Variation:** Let students put the animal name cards inside a real mitten.)
  (Supplies: Lunch sack, an optional real mitten)

- **Art Activity (Mittens):** If you have mitten die cuts, use various colors of construction paper to cut out two for each child. You can also download a pattern from online. Record the directions below on the chalkboard for students to follow.
  1. Print your name on both mittens.
  2. Cut out the mittens.
  3. Tape the yarn ends to the backs of your mittens to form a pair.
  4. Decorate your mittens.

  (Supplies: Various colors of construction paper; scissors; scotch tape; yarn pieces; glitter glue to decorate mittens with)

- **The temperature at which water freezes:** Use a demonstration thermometer to illustrate to students the temperature at which water freezes (32°).

  Purchase Fun Pops and freeze one of them. Show students a frozen and an unfrozen Popsicle and ask questions similar to those listed below.
  1. If I wanted to freeze this unfrozen Popsicle, what would I have to do? (*Put it in a freezer.*)
  2. If the freezer freezes the Popsicle, what's the temperature in the freezer? (*Below 33°*)
  3. Let's imagine that the temperature outside is at or below 32°. (It's as cold as a freezer.) Now, let's imagine that it's going to rain. Because the temperature outdoors is below 33°, what will happen to the water that would normally fall from the clouds as rain? (*It will change to ice or snow.*)
  4. When the temperature rises above 32°, what will happen to the snow that's on the ground? (*It will melt to become water.*)
  5. Why does it snow in the winter sometimes but not in the summer?

  (Supplies: Demonstration thermometer; Fun Pops, one frozen and one unfrozen)

- **Frost Experiment:** Pack a layer of ice into a fruit can. Sprinkle a layer of salt atop it. Repeat this process until the can is packed full. Shortly, frost will form on the outside of the can. While it's forming, use a freezer thermometer to take the temperature inside the can. Before determining the results, ask students to predict on their "Frost Experiment" worksheets what the temperature will be. The worksheets can be completed after the results have been determined.
  (Supplies: Ice in an ice chest & an ice scoop; fruit can; salt; freezer thermometer; a "Frost Experiment" worksheet for each child)

- **Writing/Handwriting assignment**
  Younger Students: Make a copy of the reproducible writing paper provided on page 30 and position it under a Smart Document Camera or a similar device. Discuss the correct stroke for each letter as you copy the incomplete sentence below (or one of your own choosing) onto your paper. Students should follow along with you, copying the letters on papers of their own. (This will help them to remain on task and to finish the handwriting assignment in unison.) They can then finish the sentence, add additional sentences, and include appropriate illustrations.

  My favorite thing about the winter season is...

  Older Students: Older students can write about the topic below. Let them copy the question and respond to it in their writing journals. They may also want to add appropriate illustrations.

  What is your favorite thing about the winter season and why?

Context Clues

Directions: Fill in each blank with a spelling word that makes sense in the context of the sentence. (Cross out the words from the list as you use them.) Circle the letters in the key words that make the long *o* sound (*o*, *oa*, *ow*).

Spelling words: close, froze, cold, cocoa, coat, roads, below, blow, slowly, snow, window, throw

1. Last night, the temperature fell ____________ 33°.
2. It was ____________ enough to snow.
3. When I woke up and looked out my ____________, it looked like a winter wonderland!
4. There was ____________ everywhere—even in the trees!
5. The ____________ were very slippery.
6. A decision was made to ____________ school for the day.
7. Dad still had to work, but he promised to drive ____________.
8. I put on my ____________ and went out to play.
9. I wanted to ____________ snowballs with my friends.
10. After a while, the cold wind started to ____________ harder.
11. I nearly ____________!
12. When I went back home to warm up, Mom made me hot ____________ to drink.

Spelling Word Riddles

Discussion: All spelling words this week have the long *o* sound, /ō/, and can be spelled as ***o*** as in *cold*, ***o*** with a final *e* as in *froze*, ***oa*** as in *coat*, or ***ow*** as in *snow*. The long *o* sound is the same as the name of the letter *o*. In the word *snow*, what vowel sound do you hear? (Help students hear and pronounce the long *o* sound by first saying "*o*" and then a basic word, such as *snow*.)

Directions: Reproduce the next two pages on light blue or yellow cardstock as instructed. Cut out the pieces and attach a magnetic strip to the back of each piece. Line up the yellow pieces horizontally towards the top of a magnetic board. Position the light blue strips vertically below them. (See the illustration on page 15.) Read a riddle and call on a student to identify the correct spelling word. For the first sentence, a student would identify the word *slowly* and then move the yellow magnetic strip with the letters ***ow*** between the light blue letters ***sl*** and ***ly*** to form the word ***slowly***.

1. In this word, the long *o* sound is spelled *o-w*. This is how you should drive on icy and snowy roads. (*Slowly*)
2. In this word, the long *o* sound is spelled *o*. This is the past tense of freeze. (*Froze*)
3. In this word, the long *o* sound is spelled *o-w*. You can feel the wind do this. (*Blow*)
4. In this word, the long *o* sound is spelled *o*. Schools sometimes do this after a big snowstorm. (*Close*)
5. In this word, the long *o* sound is spelled *o-w*. When I looked out of this, I saw a lot of snow outside. (*Window*)
6. In this word, the long *o* sound is spelled *o-w*. This is the opposite of *above*. (*Below*)
7. In this word, there are two long *o* sounds. It's good to drink when it is cold outside. (*Cocoa*)
8. In this word, the long *o* sound is spelled *o-a*. These become dangerous to drive on after an ice storm. (*Roads*)
9. In this word, the long *o* sound is spelled *o-w*. Rain can change to this when the temperature drops to 32° or below. (*Snow*)
10. In this word, the long *o* sound is spelled *o-a*. You wear this to keep warm in winter. (*Coat*)
11. In this word, the long *o* sound is spelled *o*. It is the antonym of hot. (*Cold*)
12. In this word, the long *o* sound is spelled *o-w*. This is something you might do with a snowball. (*Throw*)

Directions: Use light blue cardstock to reproduce this page. Cut out the pieces and attach magnetic strips to the back of each piece. (See the previous page for further instructions.)

cl	se	fr	ze
c	ld	c	c
c	t	r	ds
bel		bl	
sl	ly	sn	
wind		thr	

Directions: Use yellow cardstock to reproduce this page. Cut out the pieces and attach magnetic strips to the backs.

oa oa oa oa oa

ow ow ow ow

ow ow ow ow

o o o o o o o o

Spelling Patterns of Sounds Heard in Our Spelling Words

Directions: Fill in the missing letters that make the long *o* sound /ō/.

- Use a **blue** colored pencil for ***o***.
- Use a **red** colored pencil for ***oa***.
- Use a **green** colored pencil for ***ow***.

c l _ s **e**	c _ _ t	s l _ _ l y
f r _ z **e**	r _ _ d s	s n _ _
c _ l d	b e l _ _	w i n d _ _
c _ c _ _	b l _ _	t h r _ _

Directions: Record spelling words whose long *o* sound is spelled *o* under the **"o"** heading, words whose /ō/ sound is spelled *oa* under the **"oa"** heading, and words whose /ō/ sound is spelled *ow* under the **"ow"** heading.

o	**oa**	**ow**

Spelling Words (Long *o* /ō/ sound)

G	C	N	V	B	E	L	O	W	G	U	W
C	O	A	T	L	K	S	U	T	F	B	I
S	U	O	D	M	R	O	A	D	S	J	N
U	K	T	L	H	T	Y	V	D	X	H	D
K	P	H	K	C	N	C	O	L	D	Z	O
B	I	R	N	O	H	S	D	J	M	N	W
L	A	O	D	C	J	L	R	C	Z	S	J
O	F	W	N	O	V	O	R	K	G	N	E
W	S	L	W	A	Q	W	H	C	R	O	O
X	G	Y	Y	W	E	L	E	H	I	W	C
C	L	O	S	E	P	Y	Y	V	F	M	R
Q	Y	L	J	P	J	X	F	R	O	Z	E

Winter Words

1. clos**e**
2. froz**e**
3. cold
4. cocoa
5. coat
6. roads
7. below
8. blow
9. slowly
10. snow
11. window
12. throw

Homophones

clothes

What **clothes** should
I wear today?

Clothes are what we wear on our bodies.

close

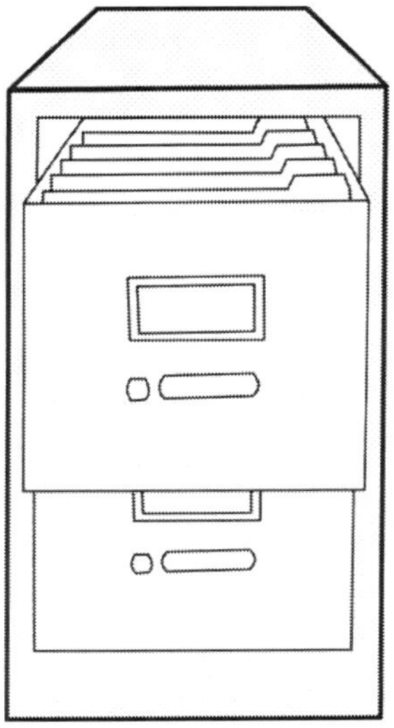

Someone forgot to
close the drawer.

Close means to shut.

Mom said, "**Close** the closet door after you pick out your **clothes**."

Underline the word that refers to what we put on our bodies.
Circle the word that means to shut.

Compound Words

Directions: When two or more root words are used together to form one word, it is called a **compound word** (snowflake). Form four compound words by adding a spelling word to the words *over*, *sill*, *storm*, and *bear*.

Spelling words: coat, polar, snow, window

over_____ _____sill _____storm _____ bear

Directions: Fill in each blank with one of the compound words that makes sense in the context of the sentence.

I was sitting on my ________________ watching the ________________ when I thought I saw a ________________ walk by! A polar bear doesn't need to wear an ________________ because its fur and fat help to keep it warm.

Compound Words

Directions: When two or more root words are used together to form one word, it is called a **compound word** (windowsill). Complete each sentence by writing a word to form a compound in the blank space. The word must make sense in the context of the sentence.

1. Last night we had a snow__________. By morning, we already had one foot of snow__________.
   (ball, cap, fall, flake, man, storm, suit)

2. Did you know that every snow__________ has a different design?
   (ball, cap, fall, flake, man, storm, suit)

3. Mandi dressed in her snow__________ and put a snow__________ on her head. Then she went outside and tossed a snow__________ at her daddy, Michael, while he was building a snow__________.
   (ball, cap, fall, flake, man, storm, suit)

Weekly Oral Language

1. Correct the asking sentence.

   is every snowflake designed differently

2. Correct the telling sentence.

   every snowflake is designed differently

3. What is the subject of the telling sentence?

   ____

4. Read the *telling sentence* again and identify the helping verb and the verb (action word).

   Helping Verb: ____________________

   Verb: ____________________

5. Write your own sentence about this week's theme.

   ____

Instruction: *Teaching Children about Capitalization, Punctuation, Apostrophe s, and Contractions (Language Book 2)* by Habakkuk Educational Materials can be purchased separately to give students the knowledge they need to complete this worksheet. After students have completed the worksheet and turned it in, complete a copy of it on your classroom screen.

Parts of Speech

Directions: Identify what parts of speech are used in the sentence below and write them in the spaces provided. Then identify the subject (naming part) and predicate (action part) of the sentence and write them in the spaces provided.

Parts of speech: Noun marker, noun, pronoun, verb, helping verb, adjective, adverb, preposition, conjunction, interjection

The	**roads**	**are**	**getting**	**slippery.**
__________	__________	__________	__________	__________

Subject (Naming Part)

____

Predicate (Action Part)

____

Instruction: *Teaching Children about Parts of Speech (Language Book 3)* by Habakkuk Educational Materials can be purchased separately to help students quickly learn the parts of speech. After students have completed this worksheet and turned it in, complete a copy of it on your classroom screen.

mouse	frog	owl
rabbit	fox	wolf
boar	bear	cricket
mole	hedgehog	badger
turtle	squirrel	lizard
racoon	skunk	snake

Scientist: ______________________________

Frost Experiment

1. Because frost formed outside the can after adding ice and salt, I predict that the temperature inside is **at or below 32°** / **above 32°**. (Circle your prediction.)

2. Temperature: ______________________________

3. My prediction was **correct** / **incorrect** because

____

____

____

Name: ______________________________

Addition and Subtraction Facts

Directions

1. Count the snowflakes above the first space and write the addend.
2. Count the snowflakes above the second space and write the addend.
3. Count all the snowflakes and write the sum after the equal sign.

________ + ________ = ________

Directions

1. Count the scarves above the first space and write the number.
2. Cross out 3 scarves.
3. Count the number of scarves left and write the answer after the equal sign.

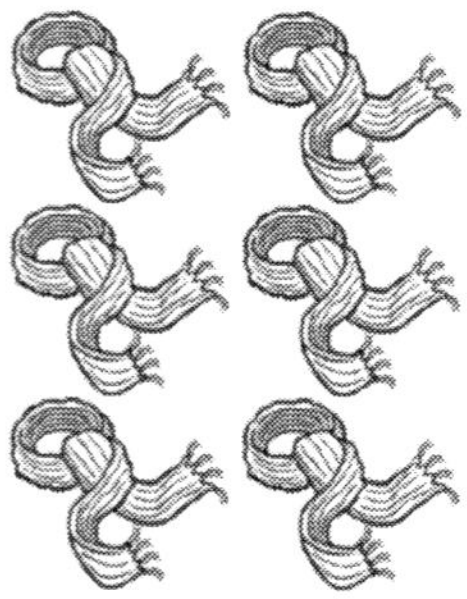

________ − ___3___ = ________

Wintery Word Problems

Directions: Cut out the pictures at the bottom of this page. Read each sentence and use the cutouts to help solve the problem. Write the addition or subtraction fact in the spaces provided.

1. It snowed 1 day in December, 5 days in January, and 3 days in February. How many days did it snow altogether?"

   _____ + _____ + _____ = _____

2. There were 6 snowflakes. Four snowflakes melted, how many snowflakes are left?

   _____ – _____ = _____ –

3. Six bears, 2 snakes, and 1 tortoise hibernated for the winter. How many animals hibernated altogether?

   _____ + _____ + _____ = _____

4. There were 5 bears. If 3 bears hibernated, how many bears are still awake?

   _____ – _____ = _____

QUESTIONS FOR THE THEMATIC UNITS GAME

Please refer to pages 28-29 of this book for instructions on how to use the unit questions.

Winter is one of the four seasons. What are the four seasons? *Lesson 18 (Science Themes)*	What happens in nature during the winter season? *Lessons 18 (Science Themes)*
What do animals like bears, bats, and skunks do in winter? *Lessons 18 (Science Themes)*	At what temperature does rain change to ice or snow? *Lessons 18 (Science Themes)*
What are the winter months? *Lessons 18 (Science Themes)*	

LESSON 18 TEST (The Winter Season)

1. What are the four seasons?

   ○ Sunday, Monday, Tuesday, and Wednesday

   ○ September, October, November, and December

   ○ Winter, spring, summer, and fall

2. What happens in nature during the winter season?

   ○ It gets very hot outside.

   ○ It can get very cold and snow.

   ○ There are a lot of tornadoes.

3. What do animals like bears, bats, and skunks do in winter?

   ○ They don't do anything differently from any other time of the year.

   ○ They hibernate.

   ○ They travel South where it is warmer.

4. At what temperature does rain change to ice or snow?

   ○ 32° or lower

   ○ Higher than 32°

   ○ 50°

5. What are the Winter months?

   ○ December, January, and February

   ○ June, July, and August

   ○ September, October, November

Note: These cards can be used for the spelling word games listed on pages 21-22.

cl<u>o</u>s**e**	fr<u>o</u>z**e**
c<u>o</u>ld	c<u>o</u>c<u>oa</u>
c<u>oa</u>t	r<u>oa</u>ds

Note: These cards can be used for the spelling word games listed on pages 21-22.

below	blow
slowly	snow
window	throw

Lesson 18_2

Optional themes for /ō/ words:
The Winter Season (18_1),
Noah's Ark (18_2)

Theme: Noah's Ark (Biblical theme)
Sound: long *o* /ō/
Spelling Patterns: *o* with final *e*, *o*, *oa*, *ow*, *oe*

Spelling Words

1. hop**e**
2. Noah
3. float
4. goat
5. boat
6. toad
7. oats
8. flowing
9. window
10. rainbow
11. crow
12. doe

Alphabetized

____
____
____
____
____
____
____
____
____
____
____
____

Inflectional endings: floats, floating, floated

Note: The initial letter of *Noah* must be capitalized.

Note: Notice that in the name *Noah*, *a* is making a vowel sound and is not silent as in the other words with *oa*.

Thematic Activities for Lesson 18
(Noah's Ark)

Thematic objectives: To familiarize students with the biblical story of Noah's Ark

Key Questions for Discussion

- What is a female deer called? (*doe or hind*)
- What is a male deer called? (*buck or stag*)

Activities, Supplies, and Preparations for this Lesson

- **Spelling Patterns of Sounds Heard in the English Language:** Display words with this week's sound on the wall display. (See pages 26-27 for details.)

- **Reproducible materials:** Prepare the reproducible materials provided with this lesson, including the "Spelling Word Riddles," the question cards, and the spelling word cards
(Supplies: Computer paper, light blue and yellow cardstock, magnetic tape)

- **Books and videos:** Read books and/or watch YouTube videos about Noah's Ark. (For any books you read, identify the title, the author, and the illustrator. Also, examine the front and back cover of the book before making any predictions about the story or discussing what it is about.)

- **Crafts:** Choose a craft activity for making rainbows from online. Decorate the room with the rainbows students make.

- **Rainbows:** Have students follow the given instructions to color the rainbow included with this lesson.
(Supplies: Red, orange, yellow, green, blue, indigo, and violet crayons or markers)

- **Noah's Ark Activity:** Follow the instructions on page 378.
(Supplies: A shoebox with a slit in the lid)

- **Writing/Handwriting assignment:** Position the handwriting assignment on the page 371 under a Smart Document Camera or a similar device. Discuss the correct stroke for each letter as you copy the sentences onto your paper. Students should follow along with you, copying the letters on papers of their own. (This will help them to remain on task and to finish the handwriting assignment in unison.)

  <u>Older Students:</u> Older students can also write about the topic below. Let them copy the question and respond to it in their writing journals. They may also want to add appropriate illustrations.

  What is your favorite part about the Noah's Ark story and why?

- **Counting by Twos Bulletin Board:** Cover the bulletin board with light blue butcher paper. Use grass border for the bottom of the board and light blue for the other sides. Use large cutout letters to title the display, "Counting by Twos." Provide a large ark shape (preferably laminated) cut from brown posterboard. On the ark, use a Sharpie to record the paraphrased verse, *Two of every kind of animal, male and female, went into the ark with Noah (Genesis 7:8-9)*. Fix the ark on the left side of the board. Provide pairs of animal die cuts walking towards the ark and several clouds at the top.

Context Clues

Directions: Fill in each blank with a spelling word that makes sense in the context of the sentence. (Cross out the words from the list as you use them.) Circle the letters in the key words that make the long *o* sound (*o*, *oa*, *ow*, *oe*).

Spelling words: hope, Noah, float, goat, boat, toad, oats, flowing, window, rainbow, crow, doe

1. God told ____________ to build an Ark.

2. Another word for *ark* is ____________.

3. Noah put his ____________ in God and built the ark.

4. As the flood waters rose, Noah's boat began to ____________.

5. The ____________ water moved the ark.

6. A male and female ____________ were safe inside the boat.

7. They liked to eat ____________.

8. There was also a male and female deer inside the ark. A female deer is called a ____________.

9. There may have even been a male and a female ____________ jumping around the ark.

10. We know that a male and a female ____________ and some other birds were there.

11. When Moses released a dove from the ____________, it later came back with an olive leaf in its mouth because the flood was almost over.

12. After the flood, God put a ____________ in the sky.

Spelling Word Riddles

Discussion: All spelling words this week have the long *o* sound, /ō/, and can be spelled as ***o*** as in *Noah*, ***o*** with a final *e* as in *hope*, ***oa*** as in *boat*, ***ow*** as in *rainbow*, or ***oe*** as in *doe*. The long *o* sound is the same as the name of the letter *o*. In the word *oats*, what vowel sound do you hear? (Help students hear and pronounce the long *o* sound by first saying "*o*" and then a basic word, such as *oats*.

Directions: Reproduce the next two pages on light blue or yellow cardstock as instructed. Cut out the pieces and attach a magnetic strip to the back of each piece. Line up the yellow pieces horizontally towards the top of a magnetic board. Position the light blue strips vertically below them. (See the illustration on page 15.) Read a riddle and call on a student to identify the correct spelling word. For the first sentence, a student would identify the word *boat* and then move the yellow magnetic strip with the letters ***oa*** between the light blue letters ***b*** and ***t*** to form the word ***boat***.

1. In this word, the long *o* sound is spelled *o-a*. This is another word for *ark*. (*Boat*)
2. In this word, the long *o* sound is spelled *o-w*. When Moses was in the ark, he used this to look outside. (*Window*)
3. In this word, the long *o* sound is spelled *o-e*. This is a female deer. (*Doe*)
4. In this word, the long *o* sound is spelled *o*. It begins with *h* and rhymes with *rope*. (*Hope*)
5. In this word, the long *o* sound is spelled *o*. God told this man to build an ark. (*Noah*)
6. In this word, the long *o* sound is spelled *o-a*. This animal is classified as an amphibian and can jump. (*Toad*)
7. In this word, the long *o* sound is spelled *o-w*. You can see water doing this in a stream. (*Flowing*)
8. In this word, the long *o* sound is spelled *o-a*. A boat can do this on water. (*Float*)
9. In this word, the long *o* sound is spelled *o-a*. Goats like to eat this. (*Oats)*
10. In this word, the long *o* sound is spelled *o-w*. This bird was in Noah's ark. (*Crow*)
11. In this word, the long *o* sound is spelled *o-a*. This is an animal that rhymes with boat. (*Goat*)
12. In this word, the long *o* sound is spelled *o-w*. God put this in the sky after the flood was over. (*Rainbow*)

Directions: Use light blue cardstock to reproduce this page. Cut out the pieces and attach magnetic strips to the back of each piece. (See the previous page for further instructions.)

h	pe	N	ah
fl	t	g	t
b	t	t	d
ts	fl	ing	
wind		rainb	
cr	d		

Directions: Use yellow cardstock to reproduce this page. Cut out the pieces and attach magnetic strips to the backs.

oa	oa	oa	oa	oa			
oa	oa	oe	oe	oe			
ow	ow	ow	ow				
o	o	o	o	o	o	o	o

QUESTIONS FOR THE THEMATIC UNITS GAME

Please refer to pages 28-29 of this book for instructions on how to use the unit questions.

What is a female deer called? *Lesson 18 (Science Themes)*	**What** is a male deer called? *Lesson 18 (Science Themes)*

Spelling Patterns of Sounds Heard in Our Spelling Words

Directions: Fill in the missing letters that make the long *o* sound /ō/.

- Use a **blue** colored pencil for ***o***.
- Use a **red** colored pencil for ***oa***.
- Use a **green** colored pencil for ***ow***.
- Use a **purple** colored pencil for ***oe***.

h _ p **e**	b _ _ t	w i n d _ _
N _ a h	t _ _ d	r a i n b _ _
f l _ _ t	_ _ t s	c r _ _
g _ _ t	f l _ _ i n g	d _ _

Directions: Record spelling words whose long *o* sound is spelled *o* under the **"o"** heading, words whose /ō/ sound is spelled *oa* under the **"oa"** heading, and so forth.

o	**oa**	**ow**	**oe**
____________	____________	____________	____________
____________	____________	____________	____________
____________	____________	____________	____________
____________	____________	____________	____________
____________	____________	____________	____________

Spelling Words (Long *o* /ō/ sound)

F	L	O	W	I	N	G	W	E	G	U	O
U	K	L	S	L	I	S	U	N	O	A	H
W	U	H	O	P	E	A	G	E	T	J	V
I	T	B	M	H	T	Y	O	D	X	H	Z
N	P	K	C	H	N	R	A	J	Z	D	R
D	I	N	F	J	D	O	T	L	M	N	A
O	D	A	B	O	A	T	W	C	C	S	I
W	F	N	Z	Q	F	H	P	K	R	T	N
X	S	L	W	T	O	A	D	C	O	W	B
O	A	T	S	W	M	E	P	H	W	Q	O
I	R	Q	E	I	P	X	Y	V	F	M	W
Q	D	O	E	B	Z	F	L	O	A	T	X

Noah's Ark

1. hop**e**
2. Noah
3. float
4. goat
5. boat
6. toad
7. oats
8. flowing
9. window
10. rainbow
11. crow
12. doe

Weekly Oral Language

1. Correct the asking sentence.

   why did God tell noah to build an ark

2. Correct the telling sentence.

   god was sending a flood

3. What is the subject of the telling sentence?

   ____

4. Read the *telling sentence* again and identify the helping verb and the verb (action word).

   Helping Verb: ______________________

   Verb: ______________________________

5. Write your own sentence about this week's theme.

   ____

Instruction: *Teaching Children about Capitalization, Punctuation, Apostrophe s, and Contractions (Language Book 2)* by Habakkuk Educational Materials can be purchased separately to give students the knowledge they need to complete this worksheet. After students have completed the worksheet and turned it in, complete a copy of it on your classroom screen.

Parts of Speech

Directions: Identify what parts of speech are used in the sentence below and write them in the spaces provided. Then identify the subject (naming part) and predicate (action part) of the sentence and write them in the spaces provided.

Parts of speech: Noun marker, noun, pronoun, verb, helping verb, adjective, adverb, preposition, conjunction, interjection

Noah	**built**	**a**	**big**	**boat.**
________	________	________	________	________

Subject (Naming Part)

____

Predicate (Action Part)

____

Instruction: *Teaching Children about Parts of Speech (Language Book 3)* by Habakkuk Educational Materials can be purchased separately to help students quickly learn the parts of speech. After students have completed this worksheet and turned it in, complete a copy of it on your classroom screen.

Handwriting directions: Copy the sentences below on the lines at the bottom of the page.

A male deer is called a buck or stag. A female deer is called a doe or hind.

buck or stag

doe or hind

Lesson: The colors of a rainbow are red, orange, yellow, green, blue, indigo, and violet. Red is the color at the top of the rainbow. The acronym used to help people remember the order of the rainbow colors is "**Roy G. Biv**." *R* stands for red, *o* for orange, *y* for yellow, *g* for green, *b* for blue, *i* for indigo, and *v* for violet.

Materials: Red, orange, yellow, green, blue, indigo, and violet crayons or markers

Directions: Use crayons or markers to put the colors of the rainbow in the correct order. Red is the color at the top of the rainbow.

Rainbow

Name: ______________________________

Addition and Subtraction Facts

Directions

1. Count the animals above the first space and write the addend.
2. Count the animals above the second space and write the addend.
3. Count all the animals and write the sum after the equal sign.

__________ + __________ = __________

Directions

1. Count the animals above the first space and write the number.
2. Cross out 2 animals.
3. Count the number of animals left and write the answer after the equal sign.

__________ − 2 = __________

Subtraction Word Problems

Directions: Color and cut out the pictures at the bottom of this page. Read each sentence and use the cutouts to help solve the problem. Write the subtraction fact in the spaces provided.

1. There were **12** animals outside the ark. If **6** animals went inside, how many animals are still outside?

   _____ – _____ = _____

2. There were **10** animals outside the ark. If **8** animals went inside, how many animals are still outside?

   _____ – _____ = _____

3. There were **12** animals outside the ark. If **10** animals went inside, how many animals are still outside?

   _____ – _____ = _____

4. There were **10** animals outside the ark. If **5** animals went inside, how many animals are still outside?

   _____ – _____ = _____

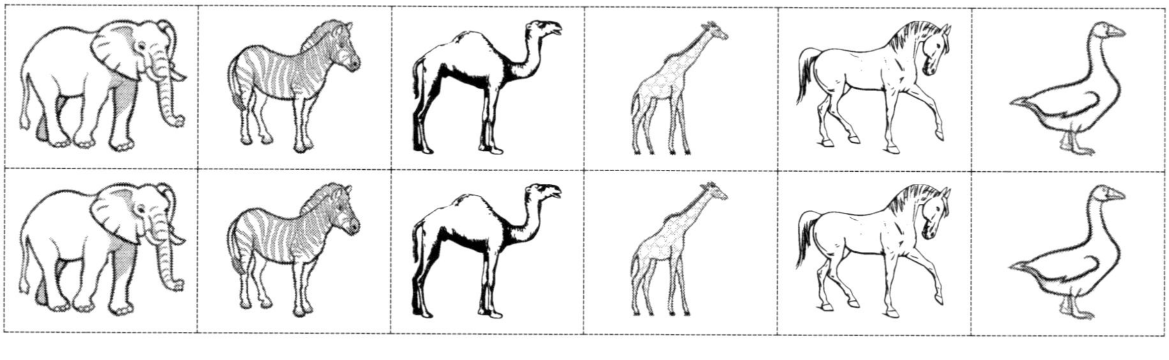

LESSON 18 TEST (Noah's Ark)

1. What is a female deer called? (Mark TWO answers.)

   ◯ Buck

   ◯ Doe

   ◯ Hind

   ◯ Stag

2. What is a male deer called? (Mark TWO answers.)

   ◯ Buck

   ◯ Doe

   ◯ Hind

   ◯ Stag

Note: These cards can be used for the spelling word games listed on pages 21-22.

hope	Noah
float	goat
boat	toad

Note: These cards can be used for the spelling word games listed on pages 21-22.

oats	flowing
window	rainbow
crow	doe

Noah's Ark Activity: This phonics activity will be used to assist students in identifying various spelling patterns of the emphasized sound. Spelling words to be studied during a theme on "Noah's Ark" contain various spelling patterns of the long *o* sound, including *oa* as in *boat*, *ow* as in *rainbow*, *oe* as in *doe*, and *o* with a final *e* as in *hope*.

Color and cut out the ark shape below and tape it to a shoebox with a slit in the lid. You will also need to cut out the spelling word cards and the word cards having the short *o* sound. Each child will choose a card to read and note if the word makes the short or long *o* sound. If it makes the long *o* sound /ō/, the child will stick it through the slit in the lid. Note that if the letter *o* in a word is followed by *a* (*oa*), *w* (*ow*), *e* (*oe*), or has a final *e* (*o-e*), it is pronounced with the long *o* sound.

Note: Flashcards on this page contain the short *o* sound rather than the long *o* sound and are used only with the "Noah's Ark Activity" described on the previous page.

dog	got
hot	long
lot	mom

Note: Flashcards on this page contain the short *o* sound rather than the long *o* sound and are used only with the "Noah's Ark Activity."

n<u>o</u>t	<u>o</u>ff
<u>o</u>n	sh<u>o</u>p
st<u>o</u>p	wr<u>o</u>ng

Lesson 19_1

Optional themes for /k/ words:
The Circus (19_1),
Transportation (19_2)

Theme: The Circus
Sound: /k/
Spelling Patterns: *c, k, ke, ck*

Spelling Words

1. cotton
2. candy
3. cats
4. cage
5. popcorn
6. circus
7. clown
8. clap
9. kangaroo
10. makeup
11. ticket
12. trick

Alphabetized

____
____
____
____
____
____
____
____
____
____
____
____

Inflectional endings: tricks, tricking, tricked

Extra challenge words: costume, character

Context Clues

Directions: Fill in each blank with a spelling word that makes sense in the context of the sentence. (Cross out the words from the list as you use them.) Circle the letters in the key words that make the /k/ sound (*c*, *k*, *ke*, *ck*).

Spelling words: cotton, candy, cats, cage, popcorn, circus, clown, clap, kangaroo, makeup, ticket, trick

1. Tonight, I went to the ____________.
2. My mom bought us both a ____________.
3. Mom got a bucket of ____________ to eat.
4. I ate something with a sugary taste that looked like ____________ on a stick.
5. They call it cotton ____________.
6. They had big ____________ like lions and tigers.
7. We weren't scared because they were in a ____________.
8. They also had a ____________ jumping around everywhere.
9. They had a super tall ____________.
10. He was wearing funny ____________ on his face.
11. He did an awesome ____________!
12. When it was over, everyone started to ____________.

Spelling Word Riddles

Discussion: All spelling words this week have the /k/ sound and can be spelled as ***c*** as in *cage*, ***k*** as in *kangaroo*, ***ke*** as in *makeup*, or ***ck*** as in *trick*. (Help students hear and pronounce the /k/ sound by first saying the sound and then a basic word, such as *cage*.)

Directions: Reproduce the next page on light blue and yellow cardstock. Cut out the pieces and attach a magnetic strip to the back of each piece. Line up the yellow pieces horizontally towards the top of a magnetic board. Position the light blue strips vertically below them. (See the illustration on page 15.) Read a riddle and call on a student to identify the correct spelling word. For the first sentence, a student would identify the word *ticket* and then move the yellow magnetic strip with the letters ***ck*** between the light blue letters ***ti*** and ***et*** to form the word ***ticket***.

1. In this word, the /k/ sound is spelled *c-k*. You can use this to get into the circus. (*Ticket*)
2. In this word, the /k/ sound is spelled *k-e*. Clowns wear this on their faces. (*Makeup*)
3. In this word, the /k/ sound is spelled *k*. This animal can jump high and has a pouch to carry its baby in. (*Kangaroo*)
4. In this word, the /k/ sound is spelled *c*. This is what a certain sugary candy on a stick looks like. (*Cotton*)
5. In this word, the /k/ sound is spelled *c*. This is something you do with your hands to show that you liked a performance. (*Clap*)
6. In this word, the /k/ sound is spelled *c*. This snack has a salty and buttery taste. (*Popcorn*)
7. In this word, the /k/ sound is spelled *c*. A circus animal is kept in this sometimes so that it doesn't escape. (*Cage*)
8. In this word, the /k/ sound is spelled *c*. This is what chocolate and other sweets are called. (*Candy*)
9. In this word, the /k/ sound is spelled *c*. This performer wears a funny wig and lots of makeup while making people laugh. (*Clown*)
10. In this word, the /k/ sound is spelled *c*. Go to this show if you want to see clowns and animals perform silly tricks. (*Circus*)
11. In this word, the /k/ sound is spelled *c-k*. Pulling a rabbit out of a hat is an example of a what? (*Trick*)
12. In this word, the /k/ sound is spelled *c*. A lion or tiger is an example of a big one. (*Cat*)

otton		andy	
ats	age	pop	orn
cir	us	lown	
angaroo		lap	tri
ma	up	ti	et

Directions: Reproduce the cards above on light blue cardstock and the cards below on yellow cardstock. Cut out the pieces and attach magnetic strips to the backs. (See the previous page for further instructions.)

c	c	c	c	c	c	c	c	c	k	k

ck	ck	ck	ke	ke	ke

Spelling Patterns of Sounds Heard in Our Spelling Words

Directions: Fill in the missing letters that make the /k/ sound.

- Use a **blue** colored pencil for ***c***.
- Use a **red** colored pencil for ***k***.
- Use a **green** colored pencil for ***ke***.
- Use a **purple** colored pencil for ***ck***.

_ o t t o n	p o p _ o r n	_ a n g a r o o
_ a n d y	c i r _ u s	m a _ _ u p
_ a t s	_ l o w n	t i _ _ e t
_ a g e	_ l a p	t r i _ _

Directions: Record spelling words whose /k/ sound is spelled *c* beside the **"c"** headings, any word whose /k/ sound is spelled *k* beside the **"k"** heading, and so forth.

c ______________________________

c ______________________________

k ______________________________

ke ______________________________

ck ______________________________

Spelling Words (/k/ sound)

G	C	M	V	B	W	C	O	T	T	O	N
P	K	A	I	C	K	S	U	T	F	B	X
O	U	K	D	A	C	A	N	D	Y	J	T
P	T	E	M	T	T	Y	V	D	X	H	R
C	P	U	C	S	N	T	D	J	C	D	I
O	I	P	F	J	D	I	A	L	I	N	C
R	A	D	Q	V	K	C	W	C	R	S	K
N	C	A	G	E	F	K	P	K	C	T	E
G	S	L	W	E	G	E	B	C	U	W	C
C	L	O	W	N	M	T	P	H	S	Q	L
I	R	Q	E	I	P	X	Y	V	F	M	A
K	A	N	G	A	R	O	O	J	P	J	P

Circus Words

1. cotton
2. candy
3. cats
4. cage
5. popcorn
6. circus
7. clown
8. clap
9. kangaroo
10. makeup
11. ticket
12. trick

Compound Words

Directions: When two or more root words are used together to form one word, it is called a **compound word** (wildcat). Form three compound words by adding a word from the word list to the words *pop*, *up*, and *cotton*.

Word list: candy, corn, make

pop________ ________up cotton ________

Directions: Fill in each blank with one of the compound words that makes sense in the context of the sentence.

We went to the circus. For snacks, Mom got a bucket of ____________________. I ate something called ____________________ that was spun around a stick. The funny clowns were wearing lots of ____________________ and doing funny tricks.

Weekly Oral Language

1. Correct the asking sentence.

   how high can a kangaroo jump

2. Correct the telling sentence.

   a kangaroo can jump 6 feet high

3. What is the subject of the telling sentence?

   ____

4. Read the *telling sentence* again and identify the helping verb and the verb (action word).

   Helping Verb: ____________________

   Verb: ____________________

5. Write your own sentence about this week's theme.

   ____

Instruction: *Teaching Children about Capitalization, Punctuation, Apostrophe s, and Contractions (Language Book 2)* by Habakkuk Educational Materials can be purchased separately to give students the knowledge they need to complete this worksheet. After students have completed the worksheet and turned it in, complete a copy of it on your classroom screen.

Parts of Speech

Directions: Identify what parts of speech are used in the sentence below and write them in the spaces provided. Then identify the subject (naming part) and predicate (action part) of the sentence and write them in the spaces provided.

Parts of speech: Noun marker, noun, pronoun, verb, helping verb, adjective, adverb, preposition, conjunction, interjection

We	went	to	the	circus!
____________	____________	____________	____________	____________

Subject (Naming Part)

____

Predicate (Action Part)

____

Instruction: *Teaching Children about Parts of Speech (Language Book 3)* by Habakkuk Educational Materials can be purchased separately to help students quickly learn the parts of speech. After students have completed this worksheet and turned it in, complete a copy of it on your classroom screen.

Spelling the /K/ Sound at the Beginning and End of a Word

To spell the /k/ sound at the beginning of a word:

- use the letter ***k*** before the letters *e*, *i*, or *y* (k**e**ep, k**i**d, K**y**le) because the letter *c* makes the /s/ sound when it's followed by *e*, *i*, or *y*;
- use the letter ***c*** before anything else (cat).

To spell the /k/ sound at the end of a word:

- use the letters ***ck*** after a short vowel (b**a**ck);
- use the letters ***ke*** after a long vowel (l**i**ke);
- use the letter ***k*** after a consonant or two vowels (wo**r**k, l**oo**k);
- use the letter ***c*** at the end of a word with two or more syllables (music).

Beginning /k/ sound	Ending /k/ sound
k______	______ck
k______	______ke
c______	______k
c______	______c

____

Words that begin with *k*: keep, key, kid, kind
Words that begin with *c*: call, came, can, canoe, car, carry, cat, caught, clean, climb, clothes, cold, color, come, cough, could, country, course, cousin, cover, cow, crayon, cry, curve, cut
Words that end with *ck*: back, black, deck, pick, block, truck
Words that end with *ke*: make, take, like, broke, joke
Words that end with *k*: ask, drink, thank, think, walk, work, break, look, took
Words that end with *c*: attic, logic, panic, public, topic

Name: ______________________________

Addition and Subtraction Facts

Directions

1. Count the clowns above the first space and write the addend.
2. Count the clowns above the second space and write the addend.
3. Count all the clowns and write the sum after the equal sign.

________ + ________ = ________

Directions

1. Count the clowns above the first space and write the number.
2. Cross out 1 clown.
3. Count the number of clowns left and write the answer after the equal sign.

________ − 1 = ________

Note: These cards can be used for the spelling word games listed on pages 21-22.

cotton	candy
cats	cage
popcorn	circus

Note: These cards can be used for the spelling word games listed on pages 21-22.

clown	clap
kangaroo	makeup
ticket	trick

Phonics Kangaroo Game: Provide a sweatshirt with a kangaroo pocket or a similar item of clothing that can be worn by a student acting as the "kangaroo." Let a student color the kangaroo in the sign below and cut it out. Provide a large piece of yarn and tape an end on both sides of the sign so that it can be worn around a child's neck.

Cut out the word cards, which include spelling words with the /k/ sound as well as words with a *c* that is pronounced as /s/ rather than /k/. (See the rule below). A student reads a word card and notes if the letter *c* is pronounced as /k/ or /s/. If it is pronounced as /k/, let him stick it in the pocket of the child representing the kangaroo.

Hard and soft *c*: The letter *c* makes both a hard /k/ and a soft /s/ sound; it makes the /s/ sound when it's followed by *e*, *i*, or *y*. (C**e**real, c**i**ty, c**y**cle)

kangaroo

Note: Flashcards on this page contain *c* pronounced as /s/ rather than /k/ and are used only with the "Kangaroo Game" described on the previous page.

certain	nice
cite	city
cycle	cymbal

Lesson 19_2

Optional themes for /k/ words:
The Circus (19_1),
Transportation (19_2)

Theme: Transportation
Sound: /k/
Spelling Patterns: *c, ck, k, ke*

Spelling Words	Alphabetized
1. car	________________
2. cab	________________
3. camel	________________
4. camper	________________
5. canoe	________________
6. carry	________________
7. cycle	________________
8. tractor	________________
9. truck	________________
10. rocket	________________
11. ark	________________
12. bike	________________

Inflectional endings: rockets, rocketing, rocketed

Bonus compound words: motorbike, motorcycle, semitruck, taxicab

Extra challenge words: bicycle, tricycle, helicopter

Thematic Activities for Lesson 19
(Transportation)

Thematic objectives: Students will learn about different forms of transportation and about the Wright Brothers, who built the first airplane in 1903.

Key Questions for Discussion
- What did Indians once use for water transportation? (*canoes*)
- Who built the first airplane in 1903? (*the Wright Brothers*)

Activities, Supplies, and Preparations for this Lesson
- **Spelling Patterns of Sounds Heard in the English Language:** Display words with this week's sound on the wall display. (See pages 26-27 for details.)

- **Reproducible materials:** Prepare the reproducible materials provided with this lesson, including the "Spelling Word Riddles," the question cards, and the spelling word cards
(<u>Supplies:</u> Computer paper, light blue and yellow cardstock, magnetic tape)

- **Books and videos:** Read books and/or watch YouTube videos about transportation and the Wright Brothers. (For any books you read, identify the title, the author, and the illustrator. Also, examine the front and back cover of the book before making any predictions about the story or discussing what it is about.)

- **Writing/Handwriting assignment:** Position one of the handwriting assignments from pages 408-409 under a Smart Document Camera or a similar device. Discuss the correct stroke for each letter as you copy the sentence onto your paper. Students should follow along with you, copying the letters on papers of their own. (This will help them to remain on task and to finish the handwriting assignment in unison.)

  <u>Older Students:</u> Older students can also write about the topic below. Let them copy the question and respond to it in their writing journals. They may also want to add appropriate illustrations.

  What is your favorite means of transportation and why?

Context Clues

Directions: Fill in each blank with a spelling word that makes sense in the context of the sentence. (Cross out the words from the list as you use them.) Circle the letters in the key words that make the /k/ sound (*c, ck, k, ke*).

Words: car, cab, camel, camper, canoe, carry, cycle, tractor, truck, rocket, ark, bike

1. Most people have a ____________ they use for transportation.

2. Some people prefer to drive a big ____________, which has an open space in the back to put things in.

3. A taxi driver can take you where you need to go in a ____________ if you don't have a car or a truck.

4. A ____________ has two wheels and is often used as a form of transportation by children.

5. When you were a baby and couldn't walk, your parents would ____________ you to your bed and to other places you needed to go.

6. A ____________ can refer to a tri<u>cycle</u>, a bi<u>cycle</u>, or a motor<u>cycle</u>.

7. A ____________ could be used to travel to space.

8. A ____________ would be a good thing to drive on a camping trip.

9. A ____________ can help farmers mow their grass, but it can also be used for transportation.

10. A __________ is an animal used for transportation in places like Egypt.

11. Indians knew how to build a ____________ for water transportation.

12. Noah built a big boat called an ____________ so that he and his family could survive the flood.

Spelling Word Riddles

Discussion: All spelling words this week have the /k/ sound and can be spelled as ***c*** as in *car*, ***ck*** as in *truck*, ***k*** as in *ark*, or ***ke*** as in *bike*. (Help students hear and pronounce the /k/ sound by first saying the sound and then a basic word, such as *car*.)

Directions: Reproduce the next page on light blue and yellow cardstock. Cut out the pieces and attach a magnetic strip to the back of each piece. Line up the yellow pieces horizontally towards the top of a magnetic board. Position the light blue strips vertically below them. (See the illustration on page 15.) Read a riddle and call on a student to identify the correct spelling word. For the first sentence, a student would identify the word *tractor* and then move the yellow magnetic strip with the letter ***c*** between the light blue letters ***tra*** and ***tor*** to form the word ***tractor***.

1. In this word, the /k/ sound is spelled *c*. This form of transportation can also be used to cut grass. (*Tractor*)
2. In this word, the /k/ sound is spelled *k*. This is a big ship that Noah built. (*Ark*)
3. In this word, the /k/ sound is spelled *c*. Indians knew how to build this for water transportation. (*Canoe*)
4. In this word, the /k/ sound is spelled *c*. Some people use this form of transportation for camping trips. (*Camper*)
5. In this word, the /k/ sound is spelled *c*. It can refer to a tricycle, a bicycle, or a motorcycle. (*Cycle*)
6. In this word, the /k/ sound is spelled *c-k*. This form of transportation can be used to travel to space. (*Rocket*)
7. In this word, the /k/ sound is spelled *c*. You can pay a taxi driver to take you somewhere in this. (*Cab*)
8. In this word, the /k/ sound is spelled *c*. This is what most grownups use for transportation. (*Car*)
9. In this word, the /k/ sound is spelled *k-e*. This has two wheels and is often used for transportation by children. (*Bike*)
10. In this word, the /k/ sound is spelled *c*. This animal is used for transportation in places like Egypt. (*Camel*)
11. In this word, the /k/ sound is spelled *c-k*. This popular form of transportation has an open space in the back to put things in. (*Truck*)
12. In this word, the /k/ sound is spelled *c*. When you were a baby, your parents would _____ you to your bed. (*carry*)

ar	ab	amel	
amper		anoe	
arry		cy	le
tra	tor	tru	ar
ro	et	bi	

Directions: Reproduce the cards above on light blue cardstock and the cards below on yellow cardstock. Cut out the pieces and attach magnetic strips to the backs. (See the previous page for further instructions.)

c	c	c	c	c	c	c	c	c	k

ck	ck	ck	ke	ke	ke

Spelling Patterns of Sounds Heard in Our Spelling Words

Directions: Fill in the missing letters that make the /k/ sound.

- Use a **blue** colored pencil for ***c***.
- Use a **purple** colored pencil for ***ck***.
- Use a **red** colored pencil for ***k***.
- Use a **green** colored pencil for ***ke***.

_ a r	_ a n o e	t r u _ _
_ a b	_ a r r y	r o _ _ e t
_ a m e l	c y _ l e	a r _
_ a m p e r	t r a _ t o r	b i _ _

Directions: Record spelling words whose /k/ sound is spelled *c* beside the **"c"** headings, words whose /k/ sound is spelled *ck* beside the **"ck"** heading, and so forth.

c ______________________________

c ______________________________

ck ______________________________

k ______________________________

ke ______________________________

Spelling Words (/k/ sound)

K	C	T	R	U	C	K	W	E	C	A	R
U	S	L	I	L	K	S	U	T	F	B	X
A	R	K	D	C	A	R	R	Y	T	J	C
C	T	B	M	H	T	Y	V	D	X	H	A
K	P	R	C	C	N	R	D	C	Z	D	M
B	I	O	F	A	D	O	A	Y	M	N	P
I	D	C	Q	M	K	C	W	C	Z	S	E
K	F	K	Z	E	F	A	P	L	G	T	R
E	S	E	W	L	G	B	B	E	R	W	O
A	G	T	Y	W	M	E	P	H	I	Q	C
I	R	Q	E	I	P	X	C	A	N	O	E
T	R	A	C	T	O	R	I	J	P	J	X

Transportation

1. car
2. cab
3. camel
4. camper
5. canoe
6. carry
7. cycle
8. tractor
9. truck
10. rocket
11. ark
12. bike

Homophones

ark	arc

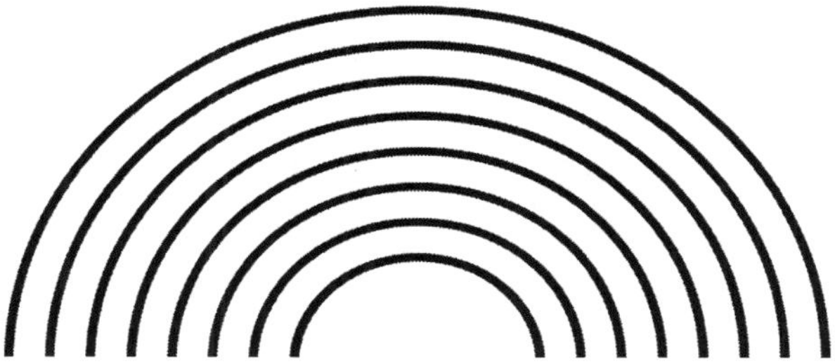

God told Noah to build an **ark** because a flood was coming.

Noah's ark was a big ship.

A rainbow looks like it is the shape of an **arc**.

An arc is a shape that looks like a half circle.

After Noah and his family left the **ark**, they saw a colorful **arc** in the sky.

Underline the word that refers to a big boat built by Noah.
Circle the word that refers to a shape that looks like a half circle.

Compound Words

Directions: When two or more root words are used together to form one word, it is called a **compound word** (motorbike). Form three compound words by adding a spelling word to the words *motor*, *semi*, and *taxi*.

Spelling words: cab, cycle, truck

motor________ semi________ taxi________

Directions: Fill in each blank with one of the compound words that makes sense in the context of the sentence.

A transportation vehicle that comes with a driver and charges a fee is a ________________. A transportation vehicle that runs with a motor but only have two wheels is a ________________. A huge truck with many wheels is called a ________________.

Weekly Oral Language

1. Correct the asking sentence.

   who built the first airplane

2. Correct the telling sentence.

   the Wright brothers built the first airplane

3. What is the subject of the telling sentence?

   ____

4. Read the *telling sentence* again and identify the verb (action word).

   ____

5. Write your own sentence about this week's theme.

   ____

Instruction: *Teaching Children about Capitalization, Punctuation, Apostrophe s, and Contractions (Language Book 2)* by Habakkuk Educational Materials can be purchased separately to give students the knowledge they need to complete this worksheet. After students have completed the worksheet and turned it in, complete a copy of it on your classroom screen.

Parts of Speech

Directions: Identify what parts of speech are used in the sentence below and write them in the spaces provided. Then identify the subject (naming part) and predicate (action part) of the sentence and write them in the spaces provided.

Parts of speech: Noun marker, noun, pronoun, verb, helping verb, adjective, adverb, preposition, conjunction, interjection

Dad	bought	a	new	truck.
__________	__________	__________	__________	__________

Subject (Naming Part)

____

Predicate (Action Part)

____

Instruction: *Teaching Children about Parts of Speech (Language Book 3)* by Habakkuk Educational Materials can be purchased separately to help students quickly learn the parts of speech. After students have completed this worksheet and turned it in, complete a copy of it on your classroom screen.

Spelling the /K/ Sound at the Beginning and End of a Word

To spell the /k/ sound at the beginning of a word:

- use the letter ***k*** before the letters *e, i,* or *y* (k**e**ep, k**i**d, K**y**le) because the letter *c* makes the /s/ sound when it's followed by *e, i,* or *y*;
- use the letter ***c*** before anything else (cat).

To spell the /k/ sound at the end of a word:

- use the letters ***ck*** after a short vowel (b**a**ck);
- use the letters ***ke*** after a long vowel (l**i**ke);
- use the letter ***k*** after a consonant or two vowels (wo**r**k, l**oo**k);
- use the letter ***c*** at the end of a word with two or more syllables (music).

Beginning /k/ sound	Ending /k/ sound
k______	______ck
k______	______ke
c______	______k
c______	______c

Words that begin with *k*: keep, key, kid, kind
Words that begin with *c*: call, came, can, canoe, car, carry, cat, caught, clean, climb, clothes, cold, color, come, cough, could, country, course, cousin, cover, cow, crayon, cry, curve, cut
Words that end with *ck*: back, black, deck, pick, block, truck
Words that end with *ke*: make, take, like, broke, joke
Words that end with *k*: ask, drink, thank, think, walk, work, break, look, took
Words that end with *c*: attic, logic, panic, public, topic

Handwriting directions: Copy the sentence on the lines.

Many Indians knew how to build canoes for water transportation.

Handwriting directions: Copy the sentence on the lines.

The Wright Brothers built the first airplane in 1903.

Name: ____________________________

Addition and Subtraction Facts

Directions

1. Count the transportation vehicles above the first space and write the addend.
2. Count the transportation vehicles above the second space and write the addend.
3. Count all the transportation vehicles and write the sum after the equal sign.

_______ + _______ = _______

Directions

1. Count the transportation vehicles above the first space and write the number.
2. Cross out 4 transportation vehicles.
3. Count the number of vehicles left and write the answer after the equal sign.

_______ − 4 = _______

QUESTIONS FOR THE THEMATIC UNITS GAME

Please refer to pages 28-29 of this book for instructions on how to use the unit questions.

What did Indians once use for water transportation? *Lesson 19 (Social Studies)*	Who built the first airplane in 1903? *Lesson 19 (Social Studies)*

LESSON 19 TEST (Transportation)

1. What did Indians once use for water transportation?

   ◯ Jet skis

   ◯ Canoes

   ◯ Sailboats

2. Who built the first airplane in 1903?

   ◯ Henry Ford

   ◯ Steve Jobs

   ◯ The Wright Brothers

Note: These cards can be used for the spelling word games listed on pages 21-22.

car	cab
camel	camper
canoe	carry

Note: These cards can be used for the spelling word games listed on pages 21-22.

cycle	tractor
truck	rocket
ark	bike

Extra Lessons for Repeated Themes and Repeated Sounds

LESSON AND THEME	SOUND	SPELLING PATTERNS	PAGE
G. The Winter Season	blends that begin with *s*	*sc, sk, sl, sn, st*	416
H. Camping Trip	short *a* /ă/	*a*	424
I. Growing Up Healthy (Community Helpers: Doctors and Dentists)	long *e* /ē/	*ea, ee, y, ey*	434
J. Farm Life, /ar/ sound	*r*-controlled vowels /ar/	*ar*	446
K. Self, I'm Special	/f/	*f, fe, gh, ph*	455
L. Fire Safety	/m/ and the *sm* blend	*m, me, mb, sm*	467

Extra Lesson A

Note to teachers: See Lesson 18_1 of Volume 1 for additional activities to use with this theme.

Theme: The Winter Season
Sounds: Blends that begin with *s*
Spelling Patterns: *sc*, *sk*, *sl*, *sn*, *st*

Spelling Words

1. scarf
2. skid
3. skate
4. sled
5. slide
6. sleet
7. slush
8. slip
9. slick
10. sneeze
11. stuck
12. frost

Alphabetized

____
____
____
____
____
____
____
____
____
____
____
____

Inflectional endings: sleets, sleeting, sleeted; slicker, slickest

Bonus compound word: ice skate

Extra challenge words: ski, sleigh, slippery, crystals, sweater

Context Clues

Directions: Fill in each blank with a spelling word that makes sense in the context of the sentence. (Cross out the words from the list as you use them.) Circle the blends that begin with *s* in each key word (*sc, sk, sl, sn, st*).

Spelling words: scarf, skid, skate, sled, slide, sleet, slush, slip, slick, sneeze, stuck, frost

1. When I woke up, there was ____________ on the windows.
2. We had snow and ____________ the night before.
3. I put on a coat and wrapped a ____________ around my neck.
4. Then I grabbed my ____________ and went outside.
5. I couldn't wait to ____________ down the hill with my friends!
6. Mom yelled out, "Be careful not to ____________!"
7. She knew the sidewalks could be ____________.
8. When I went home, Mom heard me ____________.
9. She wanted to pick me up some medicine but was afraid her tires would get ____________ in the snow.
10. I was more afraid that her car would ____________ on the ice.
11. I asked her if I could ice ____________ tomorrow if I'm not sick.
12. She said it's supposed to warm up, so I guess the ice and snow will turn to ____________.

Spelling Patterns of Sounds Heard in Our Spelling Words

Directions: Fill in the missing consonant blends of each spelling word.

- Use a **blue** colored pencil for ***sc***.
- Use a **red** colored pencil for ***sk***.
- Use a **green** colored pencil for ***sl***.
- Use a **purple** colored pencil for ***sn***.
- Use a **brown** colored pencil for ***st***.

_ _ a r f	_ _ i d e	_ _ i c k
_ _ i d	_ _ e e t	_ _ e e z e
_ _ a t e	_ _ u s h	_ _ u c k
_ _ e d	_ _ i p	f r o _ _

Directions: Record any spelling word with the *sc* blend beside the **"sc"** heading, words with the *sk* consonant blend beside the **"sk"** heading, and so forth.

sc ______________________________

sk ______________________________

sl ______________________________

sn ______________________________

st ______________________________

Spelling Words (Blends that begin with *s*)

G	S	C	A	R	F	N	A	S	L	E	D
S	K	L	I	L	K	S	U	T	F	B	X
L	U	S	D	M	S	O	F	R	O	S	T
I	T	L	M	H	T	Y	V	D	X	H	Z
C	P	I	C	H	U	R	D	J	Z	D	A
K	I	P	F	J	C	O	S	L	U	S	H
S	D	A	Q	V	K	S	W	C	Z	S	J
S	L	I	D	E	F	K	P	S	G	T	S
G	S	L	W	E	G	I	B	K	R	W	L
X	G	Y	Y	W	M	D	P	A	I	Q	E
I	R	Q	E	I	P	X	Y	T	F	M	E
Q	S	N	E	E	Z	E	I	E	P	J	T

Winter Words

1. scarf
2. skid
3. skate
4. sled
5. slide
6. sleet
7. slush
8. slip
9. slick
10. sneeze
11. stuck
12. frost

Weekly Oral Language

1. Correct the asking sentence.

   can you ice skate

2. Correct the telling sentence.

   i can learn how to ice skate

3. What is the subject of the telling sentence?

   ____

4. Read the *telling sentence* again and identify the helping verb and the verb (action word).

   Helping Verb: ______________________

   Verb: ______________________________

5. Write your own sentence about this week's theme.

   ____

Instruction: *Teaching Children about Capitalization, Punctuation, Apostrophe s, and Contractions (Language Book 2)* by Habakkuk Educational Materials can be purchased separately to give students the knowledge they need to complete this worksheet. After students have completed the worksheet and turned it in, complete a copy of it on your classroom screen.

Parts of Speech

Directions: Identify what parts of speech are used in the sentence below and write them in the spaces provided. Then identify the subject (naming part) and predicate (action part) of the sentence and write them in the spaces provided.

Parts of speech: Noun marker, noun, pronoun, verb, helping verb, adjective, adverb, preposition, conjunction, interjection

I have a purple sled.

__________ __________ __________ __________ __________

Subject (Naming Part)

____

Predicate (Action Part)

____

Instruction: *Teaching Children about Parts of Speech (Language Book 3)* by Habakkuk Educational Materials can be purchased separately to help students quickly learn the parts of speech. After students have completed this worksheet and turned it in, complete a copy of it on your classroom screen.

Note: These cards can be used for the spelling word games listed on pages 21-22.

scarf	skid
skate	sled
slide	sleet

Note: These cards can be used for the spelling word games listed on pages 21-22.

slush	slip
slick	sneeze
stuck	frost

Extra Lesson B

Theme: Camping Trip
Sound: Short *a* /ă/
Spelling Pattern: *a*

Spelling Words	Alphabetized
1. back	______________
2. pack	______________
3. bag	______________
4. flashlight	______________
5. bats	______________
6. map	______________
7. racoon	______________
8. napkins	______________
9. camp	______________
10. camper	______________
11. lantern	______________
12. canteen	______________

Inflectional endings: camps, camping, camped

Bonus compound words: backpack, campfire, campground, campsite, sleeping bag

Context Clues

Directions: Fill in each blank with a spelling word that makes sense in the context of the sentence. (Cross out the words from the list as you use them.) Circle the letter in the key words that makes the short *a* sound /ă/.

Spelling words: back, pack, bag, flashlight, bats, map, racoon, napkins, camp, camper, lantern, canteen

1. I am going to a ____________ next week.
2. It will be fun being a ____________!
3. I might get to sleep outside in my sleeping ____________.
4. That way I can watch the ____________ fly by!
5. I could even see a ____________ looking around for food.
6. I have to ____________ everything I'll need.
7. Mom bought me a pack to carry on my ____________.
8. I'll pack a _______________ so I can see in the dark.
9. Some campers will probably bring a ____________ instead of a flashlight.
10. I'll pack a ____________ filled with water for when I get thirsty.
11. I'll also bring some ____________ to clean my hands with.
12. Then, when I get to camp, I'll get a ____________ so that I can find my way around the campground.

Spelling Words (Short *a* /ă/ sound)

G	C	A	N	T	E	E	N	E	N	U	F
U	K	B	I	L	K	S	U	T	A	B	L
C	U	A	D	M	T	E	O	A	P	J	A
A	T	C	Z	B	A	G	V	D	K	H	S
M	P	K	C	H	N	R	D	J	I	D	H
P	I	P	F	J	C	O	A	L	N	N	L
E	A	A	Q	V	A	M	W	C	S	S	I
R	F	C	Z	Q	M	H	P	K	G	T	G
G	S	K	W	E	P	H	B	A	T	S	H
L	A	N	T	E	R	N	P	H	I	Q	T
I	R	Q	E	I	P	X	Y	V	F	M	R
Q	M	A	P	B	Z	R	A	C	O	O	N

Camping Trip

1. back
2. pack
3. bag
4. flashlight
5. bats
6. map
7. racoon
8. napkins
9. camp
10. camper
11. lantern
12. canteen

Compound Words

Directions: When two or more root words are used together to form one word, it is called a **compound word** (camp̲site). Form four compound words by adding a spelling word to the words *back*, *fire*, *light*, and *sleeping*.

Spelling words: bag, camp, flash, pack

_____light _____fire back_____ sleeping _____

Directions: Fill in each blank with one of the compound words that makes sense in the context of the sentence.

Today, Mike is going on a camping trip. When he arrives at the campsite, he will unload his ________________. For dinner, he will roast hotdogs over a ________________. When the sun goes down, he will turn on his ________________. When he gets tired, he will crawl into his ________________.

Weekly Oral Language

1. Correct the asking sentence.

   what should i bring to camp

2. Correct the telling sentence.

   you should bring your sleeping bag

3. What is the subject of the telling sentence?

   ____

4. Read the *telling sentence* again and identify the helping verb and the verb (action word).

   Helping Verb: ______________________

   Verb: ______________________________

5. Write your own sentence about this week's theme.

   ____

Instruction: *Teaching Children about Capitalization, Punctuation, Apostrophe s, and Contractions (Language Book 2)* by Habakkuk Educational Materials can be purchased separately to give students the knowledge they need to complete this worksheet. After students have completed the worksheet and turned it in, complete a copy of it on your classroom screen.

Parts of Speech

Directions: Identify what parts of speech are used in the sentence below and write them in the spaces provided. Then identify the subject (naming part) and predicate (action part) of the sentence and write them in the spaces provided.

Parts of speech: Noun marker, noun, pronoun, verb, helping verb, adjective, adverb, preposition, conjunction, interjection

They	**are**	**going**	**to**	**camp.**
__________	__________	__________	__________	__________

Subject (Naming Part)

____

Predicate (Action Part)

____

Instruction: *Teaching Children about Parts of Speech (Language Book 3)* by Habakkuk Educational Materials can be purchased separately to help students quickly learn the parts of speech. After students have completed this worksheet and turned it in, complete a copy of it on your classroom screen.

Name: ______________________________

Addition and Subtraction Facts

Directions

1. Count the tents above the first space and write the addend.
2. Count the tents above the second space and write the addend.
3. Count all the tents and write the sum after the equal sign.

__________ + __________ = __________

Directions

1. Count the tents above the first space and write the number.
2. Cross out 4 tents.
3. Count the number of tents left and write the answer after the equal sign.

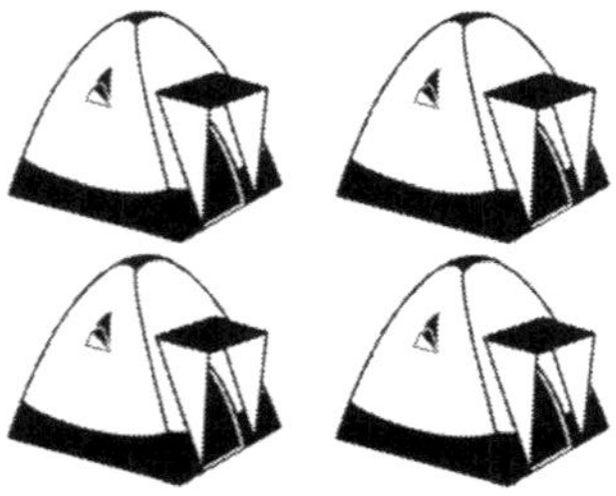

__________ − 4 = __________

Name: ______________________________

Addition and Subtraction Word Problems

Directions: Color and cut out the pictures at the bottom of this page. Read each sentence and use the cutouts to help solve the problem. Write the addition or subtraction fact in the spaces provided.

4. A fisherman caught **6** fish. He let the smallest one go. How many fish does he have left?

   _____ − _____ = _____

5. A fisherman caught **8** fish. His son caught **2**. How many fish did they catch altogether?

   _____ + _____ = _____

6. A fisherman caught **10** fish. He ate **3** of them for dinner. How many fish does he have left?

   _____ − _____ = _____

FISHING GAME: Cut out the cards and attach a paperclip to each word. Prepare a magnetic fishing pole by tying one end of a string of yarn to a stick or dowel rod and the other to a magnet with a hole in it. Let each child "fish" for a word and instruct him/her to name the letters of a caught word before reading it.

back	pack
bag	flashlight
bats	map

Note: The cards on pages 77-78 can be used for the spelling word games listed on pages 21-22.

r<u>a</u>coon	n<u>a</u>pkins
c<u>a</u>mp	c<u>a</u>mper
l<u>a</u>ntern	c<u>a</u>nteen

Extra Lesson C

Note to teachers: See Lesson 21 of Volume 2 for additional activities to use with this theme.

Theme: Growing Up Healthy (Community Helpers: Doctors and Dentists)
Sound: long *e* /ē/
Spelling Patterns: *ea*, *ee*, *y*, *ey*

Spelling Words	Alphabetized
1. eat	______________
2. clean	______________
3. peas	______________
4. peach	______________
5. meals	______________
6. meat	______________
7. teeth	______________
8. sleep	______________
9. green	______________
10. dairy	______________
11. healthy	______________
12. honey	______________

Inflectional endings: cleans, cleaning, cleaned

Extra challenge words: athlete, celery, protein

Context Clues

Directions: Fill in each blank with a spelling word that makes sense in the context of the sentence. (Cross out the words from the list as you use them.) Circle the letters in the key words that make the long *e* sound (*ea*, *ee*, *y*, *ey*).

Spelling words: eat, clean, peas, peach, meals, meat, teeth, sleep, green, dairy, healthy, honey

1. My parents want me to grow up ____________.
2. That's why my mom prepares ____________ that are good for us.
3. I always ____________ a healthy breakfast.
4. If I want something sweet, I get to put ____________ on my biscuit.
5. After school, she gives me a ____________ or some other fruit to snack on.
6. I also get a glass of milk from the ____________ food group.
7. For dinner, we usually have ____________ vegetables.
8. My favorite vegetable is ____________.
9. We also eat beef or another kind of ____________.
10. I always ____________ my hands before I eat.
11. After dinner, I brush my ____________.
12. Then I crawl into bed so that I can get a good night's ____________.

Spelling Word Riddles

Discussion: All spelling words this week have the long *e* sound, /ē/, and can be spelled as ***ea*** as in *clean*, ***ee*** as in *sleep*, ***y*** as in *baby*, or ***ey*** as in *honey*. The long *e* sound is the same as the name of the letter *e*. In the word *eat*, what vowel sound do you hear? (Help students hear and pronounce the long *e* sound by first saying "*e*" and then a basic word, such as *eat*.)

Directions: Reproduce the next two pages on light blue or yellow cardstock as instructed. Cut out the pieces and attach a magnetic strip to the back of each piece. Line up the yellow pieces horizontally towards the top of a magnetic board. Position the light blue strips vertically below them. (See the illustration on page 15.) Read a riddle and call on a student to identify the correct spelling word. For the first sentence, a student would identify the word *clean* and then move the yellow magnetic strip with the letters ***ea*** between the light blue letters ***cl*** and ***n*** to form the word ***clean***.

1. In this word, the long *e* sound is spelled *e-a*. This is what your hands should be before you eat. (*Clean*)
2. In this word, the long *e* sound is spelled *e-e*. This is what you do in your bed at night. (*Sleep*)
3. In this word, the long *e* sound is spelled *e-a*. This is a fruit and also the name of a color. (*Peach*)
4. In this word, the long *e* sound is spelled *y*. This is the food group that milk and ice cream belong in. (*Dairy*)
5. In this word, the long *e* sound is spelled *e-a*. Many people eat these three times every day. (*Meals*)
6. In this word, the long *e* sound is spelled *y*. If you want to be this, you should exercise and eat foods that are good for you. (*Healthy*)
7. In this word, the long *e* sound is spelled *e-a*. This is a word for foods like roast, steak, and chicken. (*Meat*)
8. In this word, the long *e* sound is spelled *e-e*. This is the color of many plants. (*Green*)
9. In this word, the long *e* sound is spelled *e-y*. This is something sweet that is made by bees. (*Honey*)
10. In this word, the long *e* sound is spelled *e-a*. This is a green vegetable. (*Peas*)
11. In this word, the long *e* sound is spelled *e-a*. This is what you do with food. (*Eat*)
12. In this word, the long *e* sound is spelled *e-e*. This is what you chew your food with. (*Teeth*)

Directions: Use light blue cardstock to reproduce this page. Cut out the pieces and attach magnetic strips to the back of each piece. (See the previous page for further instructions.)

t	cl	n	
p	s	p	ch
m	ls	m	t
t	th	sl	p
gr	n	dair	
health		hon	

Directions: Use yellow cardstock to reproduce the letters. Cut out the letter pieces and attach magnetic strips to the backs.

ea	ea	ea	ea	ea
ea	ee	ee	ee	ee
y	y	y	ey	ey

Spelling Patterns of Sounds Heard in Our Spelling Words

Directions: Fill in the missing letters that make the long *e* sound /ē/.

- Use a **blue** colored pencil for ***ea***.
- Use a **red** colored pencil for ***ee***.
- Use a **green** colored pencil for ***y***.
- Use a **purple** colored pencil for ***ey***.

_ _ t	m _ _ l s	g r _ _ n
c l _ _ n	m _ _ t	d a i r _
p _ _ s	t _ _ t h	h e a l t h _
p _ _ c h	s l _ _ p	h o n _ _

Directions: Record spelling words whose long *e* sound is spelled *ea* under the **"ea"** heading, words whose /ē/ sound is spelled *ee* under the **"ee"** heading, and so forth.

ea	**ee**	**y**	**ey**
____________	____________	____________	____________
____________	____________	____________	____________
____________	____________	____________	____________
____________	____________	____________	____________
____________	____________	____________	____________
____________	____________	____________	____________

Spelling Words (Long *e* /ē/ sound)

G	C	T	E	E	T	H	N	E	G	U	O
U	K	L	I	L	T	U	S	L	E	E	P
G	R	E	E	N	A	R	O	E	T	J	V
C	T	B	M	H	T	D	A	I	R	Y	Z
K	P	P	C	H	H	R	D	J	Z	D	H
B	I	E	F	J	O	O	E	A	T	N	E
M	A	A	Q	V	N	M	Y	C	Z	S	A
E	F	C	Z	Q	E	H	P	K	M	T	L
A	S	H	W	E	Y	H	E	C	E	W	T
L	G	Y	Y	W	M	E	A	H	A	Q	H
S	R	P	L	M	P	X	S	V	T	M	Y
Q	Y	L	F	C	L	E	A	N	P	J	X

Growing Up Healthy

1. eat
2. clean
3. peas
4. peach
5. meals
6. meat
7. teeth
8. sleep
9. green
10. dairy
11. healthy
12. honey

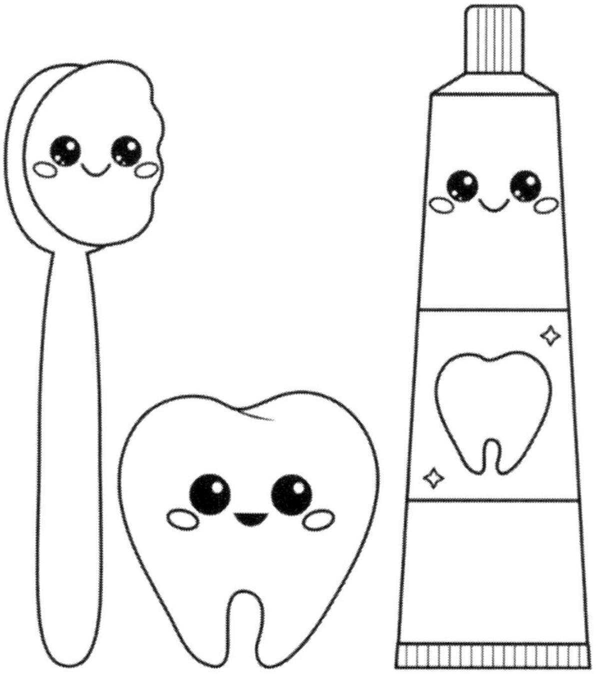

Homophones

meat

Turkey is a favorite holiday **meat**.

Meat is a food.

meet

It's so nice to **meet** you!

Meet is coming to know a person or getting together with them.

Let's **meet** at the **meat** shop to buy some steak for dinner.

Underline the word that refers to a food.
Circle the word that refers to coming to know a person or getting together with them.

Weekly Oral Language

1. Correct the asking sentence.

   how much have you grown since last year

2. Correct the telling sentence.

   i have grown two inches

3. What is the subject of the telling sentence?

   ____

4. Read the *telling sentence* again and identify the helping verb and the verb (action word).

   Helping Verb: ____________________

   Verb: ____________________________

5. Write your own sentence about this week's theme.

   ____

Instruction: *Teaching Children about Capitalization, Punctuation, Apostrophe s, and Contractions (Language Book 2)* by Habakkuk Educational Materials can be purchased separately to give students the knowledge they need to complete this worksheet. After students have completed the worksheet and turned it in, complete a copy of it on your classroom screen.

Parts of Speech

Directions: Identify what parts of speech are used in the sentence below and write them in the spaces provided. Then identify the subject (naming part) and predicate (action part) of the sentence and write them in the spaces provided.

Parts of speech: Noun marker, noun, pronoun, verb, helping verb, adjective, adverb, preposition, conjunction, interjection

Peaches	**and**	**berries**	**are**	**fruits.**
__________	__________	__________	__________	__________

Subject (Naming Part)

____

Predicate (Action Part)

____

Instruction: *Teaching Children about Parts of Speech (Language Book 3)* by Habakkuk Educational Materials can be purchased separately to help students quickly learn the parts of speech. After students have completed this worksheet and turned it in, complete a copy of it on your classroom screen.

Note: These cards can be used for the spelling word games listed on pages 21-22.

eat	clean
peas	peach
meals	meat

Note: These cards can be used for the spelling word games listed on pages 21-22.

teeth	sleep
green	dairy
healthy	honey

Extra Lesson D

Theme: Farm Life
Sound: /ar/
Spelling Pattern: *ar*

Note to teachers: See Lesson 12 of Volume 1 for additional activities to use with this theme.

Spelling Words	Alphabetized
1. hard	____________
2. bark	____________
3. farm	____________
4. farmer	____________
5. stars	____________
6. start	____________
7. barn	____________
8. barley	____________
9. harvest	____________
10. far	____________
11. yard	____________
12. jars	____________

Inflectional endings: barks, barking, barked; harder, hardest

Bonus compound words: farmyard, barnyard

Context Clues

Directions: Fill in each blank with a spelling word that makes sense in the context of the sentence. (Cross out the words from the list as you use them.) Circle the letters in the key words that make the /ar/ sound.

Spelling words: hard, bark, farm, farmer, stars, start, barn, barley, harvest, far, yard, jars

1. Oliver is a ____________.
2. He lives on a ____________ with his wife and dog.
3. Their home is ____________ away from other houses.
4. The couple works very ____________.
5. They must ____________ their chores early in the morning.
6. Farmer Oliver will ____________ his barley today.
7. He uses the ____________ to feed his cows.
8. The cows live in the ____________.
9. His wife will prepare different foods to store into ____________.
10. Their dog is in the ____________.
11. He thinks it's his job to ____________ at all the critters.
12. After supper, they like to sit on the porch and watch the ____________ in the sky.

Spelling Words (/ar/ sound)

F	A	R	Z	H	A	R	V	E	S	T	B
U	K	L	I	J	K	S	U	T	F	B	F
S	U	Y	D	B	O	E	T	A	B	J	A
C	T	A	M	A	T	H	V	D	A	H	R
S	P	R	C	R	N	A	D	J	R	D	M
T	I	D	F	N	D	R	A	L	K	N	E
A	A	D	Q	V	K	D	W	C	Z	S	R
R	F	R	X	Z	M	H	P	K	G	T	E
S	S	L	W	E	S	T	A	R	T	G	J
X	F	A	R	M	M	E	P	H	I	Q	A
I	R	Q	E	I	P	X	Y	V	F	M	R
Q	Y	B	A	R	L	E	Y	J	P	J	S

Farm Life Words

1. hard
2. bark
3. farm
4. farmer
5. stars
6. start
7. barn
8. barley
9. harvest
10. far
11. yard
12. jars

Weekly Oral Language

1. Correct the asking sentence.

   what is the dog barking at

2. Correct the telling sentence.

   the dog is barking at a rabbit

3. What is the subject of the telling sentence?

   ____

4. Read the *telling sentence* again and identify the helping verb and the verb (action word).

   Helping Verb: ____________________

   Verb: ____________________

5. Write your own sentence about this week's theme.

   ____

Instruction: *Teaching Children about Capitalization, Punctuation, Apostrophe s, and Contractions (Language Book 2)* by Habakkuk Educational Materials can be purchased separately to give students the knowledge they need to complete this worksheet. After students have completed the worksheet and turned it in, complete a copy of it on your classroom screen.

Parts of Speech

Directions: Identify what parts of speech are used in the sentence below and write them in the spaces provided. Then identify the subject (naming part) and predicate (action part) of the sentence and write them in the spaces provided.

Parts of speech: Noun marker, noun, pronoun, verb, helping verb, adjective, adverb, preposition, conjunction, interjection

The	**farmer**	**is**	**working**	**hard.**
__________	__________	__________	__________	__________

Subject (Naming Part)

____

Predicate (Action Part)

____

Instruction: *Teaching Children about Parts of Speech (Language Book 3)* by Habakkuk Educational Materials can be purchased separately to help students quickly learn the parts of speech. After students have completed this worksheet and turned it in, complete a copy of it on your classroom screen.

Note: These cards can be used for the spelling word games listed on pages 21-22.

h<u>ar</u>d	b<u>ar</u>k
f<u>ar</u>m	f<u>ar</u>mer
st<u>ar</u>s	st<u>ar</u>t

Note: These cards can be used for the spelling word games listed on pages 21-22.

b<u>ar</u>n	b<u>ar</u>ley
h<u>ar</u>vest	f<u>ar</u>
y<u>ar</u>d	j<u>ar</u>s

Phonics Farm Game: Reproduce the barn below and tape it to a lunch bag or to a shoebox with a slit in the lid. Cut out the spelling word cards as well as the word cards having *ar* that makes the /ər/ sound as in *molar*. Each child will choose a card to read and note if the word makes the /ar/ or /ər/ sound. If it makes the /ar/ sound, the child will stick it in the bag or through the slit in the lid. (Word cards representing both the /ar/ and /ər/ sounds are provided, but only those with the /ar/ sound will be placed in the box.)

Rule: The letter combinations *ar* and *or* make the /ər/ sound when they end a word with more than one syllable. (Molar, color)

Note: Flashcards on this page contain *ar* pronounced as /ər/ rather than /ar/ and are used only with the "Farm Game" described on the previous page.

cellar	collar
dollar	molar
polar	solar

Extra Lesson E

Theme: Self, I'm Special
Sound: /f/
Spelling Patterns: *f, fe, gh ph*

Spelling Words	Alphabetized
1. family	____________
2. face	____________
3. features	____________
4. foot	____________
5. feet	____________
6. fingers	____________
7. self	____________
8. myself	____________
9. future	____________
10. life	____________
11. laugh	____________
12. photo	____________

Inflectional endings: laughs, laughing, laughed

Bonus compound words: fingerprint, footprint, lifetime, photograph, yourself

Extra challenge words: favor, favorite, useful

Context Clues

Directions: Fill in each blank with a spelling word that makes sense in the context of the sentence. (Cross out the words from the list as you use them.) Circle the letters in the key words that make the /f/ sound (*f*, *fe*, *gh*, *ph*).

Spelling words: family, face, features, foot, feet, fingers, self, myself, future, life, laugh, photo

1. ____________ refers to who I am.
2. When I look at ____________ in the mirror, I don't look like you.
3. When I look at myself in a ____________, I don't look like you.
4. Even though I have two eyes, a nose, and a mouth on my ____________, they don't look just like yours.
5. That's because everyone has different facial ____________.
6. The ____________ on our hands don't make the same prints.
7. The ____________ we walk with may not be the same size.
8. My ____________ won't even fit inside my little sister's shoe!
9. When we ____________ at jokes, we don't sound exactly alike.
10. Unless we our related, we have a different ____________.
11. My life in the ____________ will not be just like yours.
12. God has a plan for my ____________ and a plan for yours.

Spelling Word Riddles

Discussion: All spelling words this week have the /f/ sound and can be spelled as ***f*** as in *face*, ***fe*** in *life*, ***gh*** in *laugh*, or ***ph*** as in *photo*. (Help students hear and pronounce the /f/ sound by first saying the sound and then a basic word, such as *ace*.)

Directions: Reproduce the next two pages on light blue or yellow cardstock as instructed. Cut out the pieces and attach a magnetic strip to the back of each piece. Line up the yellow pieces horizontally towards the top of a magnetic board. Position the light blue strips vertically below them. (See the illustration on page 15.) Read a riddle and call on a student to identify the correct spelling word. For the first sentence, a student would identify the word *fingers* and then move the yellow magnetic strip with the letter ***f*** before the light blue letters ***ingers*** to form the word ***fingers***.

1. In this word, the /f/ sound is spelled *f*. You have these on your hands. (*Fingers*)
2. In this word, the /f/ sound is spelled *f*. A clown wears makeup on this. (*Face*)
3. In this word, the /f/ sound is spelled *p-h*. This is another word for *picture*. (*Photo*)
4. In this word, the /f/ sound is spelled *f*. This word rhymes with *teachers*. (*Features*)
5. In this word, the /f/ sound is spelled *f*. This is the singular form of *feet*. (*Foot*)
6. In this word, the /f/ sound is spelled *f*. This is a compound word. (*Myself*)
7. In this word, the /f/ sound is spelled *f-e*. This is what you live. (*Life*)
8. In this word, the /f/ sound is spelled *f*. This word begins with *s* and it is who you see when you look in the mirror. (*Self*)
9. In this word, the /f/ sound is spelled *f*. This is a group of people who may live in the same house, like a father, mother, brother, and sister. (*Family*)
10. In this word, the /f/ sound is spelled *g-h*. You may do this when you hear a joke. (*Laugh*)
11. In this word, the /f/ sound is spelled *f*. It is the opposite of *past*. (*Future*)
12. In this word, the /f/ sound is spelled *f*. These help you to walk. (*Feet*)

Directions: Use light blue cardstock to reproduce this page. Cut out the pieces and attach magnetic strips to the back of each piece. (See the previous page for further instructions.)

amily	ace	oot
eatures	eet	sel
ingers	mysel	
uture	li	lau
oto		

Directions: Use yellow cardstock to reproduce this page. Cut out the pieces and attach magnetic strips to the backs.

f	f	f	f	f	f	f	f	f	f	f
fe	fe	fe	fe	fe	fe	fe				
gh	gh	ph	ph	ph						

Spelling Patterns of Sounds Heard in Our Spelling Words

Directions: Fill in the missing letters that make the /f/ sound.

- Use a **blue** colored pencil for ***f***.
- Use a **red** colored pencil for ***fe***.
- Use a **green** colored pencil for ***gh***.
- Use a **purple** colored pencil for ***ph***.

_ a m i l y	_ e e t	_ u t u r e
_ a c e	_ i n g e r s	l i _ _
_ e a t u r e s	s e l _	l a u _ _
_ o o t	m y s e l _	_ _ o t o

Directions: Record spelling words whose /f/ sound is spelled *f* beside the **"f"** headings, any word whose /f/ sound is spelled *fe* beside the **"fe"** heading, any word whose /f/ sound is spelled *gh* beside the **"gh"** heading, and any word whose /f/ sound is spelled *ph* beside the **"ph"** heading.

f ______________________________

f ______________________________

fe ______________________________

gh ______________________________

ph ______________________________

Spelling Words (/f/ sound)

F	E	A	T	U	R	E	S	E	G	M	O
U	S	L	I	L	K	S	U	T	F	Y	X
T	U	F	I	N	G	E	R	S	T	S	V
P	T	B	M	H	T	Y	V	D	X	E	Z
H	P	L	A	U	G	H	D	F	Z	L	A
O	I	N	F	J	D	O	A	A	M	F	L
T	D	F	L	I	F	E	W	C	Z	S	J
O	F	U	Z	Q	F	H	P	E	G	T	F
F	S	T	W	F	G	H	B	C	R	W	O
E	G	U	Y	E	M	S	E	L	F	Q	O
I	R	R	E	E	P	X	Y	V	F	M	T
Q	Y	E	F	T	F	A	M	I	L	Y	X

Self, I'm Special

1. family
2. face
3. features
4. foot
5. feet
6. fingers
7. self
8. myself
9. future
10. life
11. laugh
12. photo

Compound Words

Directions: When two or more root words are used together to form one word, it is called a **compound word** (<u>child</u>hood). Form six compound words by adding a spelling word to the words *my*, *your*, *time*, *prints*, and *graph*.

Spelling words: finger, foot, life, photo, self

my________ your________ ________time

________prints ________prints ________graph

Directions: Fill in each blank with one of the compound words that makes sense in the context of the sentence.

This is a ________________ of me when I was a newborn in the hospital. I like this baby picture of ________________. They even took ________________ of my tiny feet. Did you know that your footprints and ________________ are different from everyone else's? Your fingers and feet get bigger during childhood, but your prints stay the same throughout your ________________. Do you have a newborn picture of ________________ to show me?

Weekly Oral Language

1. Correct the asking sentence.

   what is your favorite color

2. Correct the telling sentence.

   my favorite color is blue

3. What is the subject of the telling sentence?

   ____

4. Read the *telling sentence* again and identify the verb (action word).

   ____

5. Write your own sentence about this week's theme.

   ____

Instruction: *Teaching Children about Capitalization, Punctuation, Apostrophe s, and Contractions (Language Book 2)* by Habakkuk Educational Materials can be purchased separately to give students the knowledge they need to complete this worksheet. After students have completed the worksheet and turned it in, complete a copy of it on your classroom screen.

Parts of Speech

Directions: Identify what parts of speech are used in the sentence below and write them in the spaces provided. Then identify the subject (naming part) and predicate (action part) of the sentence and write them in the spaces provided.

Parts of speech: Noun marker, noun, pronoun, verb, helping verb, adjective, adverb, preposition, conjunction, interjection

My	fingers	make	unique	prints.
__________	__________	__________	__________	__________

Subject (Naming Part)

____

Predicate (Action Part)

____

Instruction: *Teaching Children about Parts of Speech (Language Book 3)* by Habakkuk Educational Materials can be purchased separately to help students quickly learn the parts of speech. After students have completed this worksheet and turned it in, complete a copy of it on your classroom screen.

Note: These cards can be used for the spelling word games listed on pages 21-22.

family	face
features	foot
feet	fingers

Note: These cards can be used for the spelling word games listed on pages 21-22.

self	myself
future	life
laugh	photo

Extra Lesson F

Theme: Fire Safety (Community Helpers: Firemen)
Sound: /m/ and *sm* blend
Spelling Patterns: *m*, *me*, *mb*, and *sm*

Note to teachers: See Lesson 7 of Volume 1 for additional activities to use with this theme.

Spelling Words	Alphabetized
1. fireman	____________
2. firemen	____________
3. mask	____________
4. match	____________
5. helmet	____________
6. alarm	____________
7. damage	____________
8. home	____________
9. flame	____________
10. climb	____________
11. smell	____________
12. smoke	____________

Inflectional endings: alarms, alarming, alarmed

Extra challenge word: smother

Context Clues

Directions: Fill in each blank with a spelling word that makes sense in the context of the sentence. (Cross out the words from the list as you use them.) Circle the letters in the key words that make the /m/ sound (*m*, *me*, *mb*) or the *sm* consonant blend.

Spelling words: fireman, firemen, mask, match, helmet, alarm, damage, home, flame, climb, smell, smoke

1. Do you have a smoke ____________ in your house?

2. Every ____________ should have one.

3. The alarm will go off if it detects ____________.

4. If you hear the alarm or ____________ smoke, go outside and call 911.

5. ____________ will come to help you.

6. They may have to ____________ a ladder to help them fight the fire.

7. A fireman may look a little scary with his ____________ on, but it helps him to be able to breathe through the smoke.

8. A fireman also wears a ____________ on his head.

9. A fireman told my class that a child should never light a ____________.

10. It could start a fire and cause a lot of ____________.

11. He also gave us red flannel shaped like a ____________ to put on our clothes, and the flame fell off when we stopped, dropped, and rolled.

12. The ____________ said that you can smother the fire when you stop, drop, and roll because fire needs oxygen to keep burning.

Spelling Patterns of Sounds Heard in Our Spelling Words

Directions: Fill in the missing letters that make the /m/ sound or that have the *sm* consonant blend.

- Use a **blue** colored pencil for ***m***.
- Use a **red** colored pencil for ***me***.
- Use a **green** colored pencil for ***mb***.
- Use a **purple** colored pencil for the consonant blend ***sm***.

f i r e _ a n	h e l _ e t	f l a _ _
f i r e _ e n	a l a r _	c l i _ _
_ a s k	d a _ a g e	_ _ e l l
_ a t c h	h o _ _	_ _ o k e

Directions: Record spelling words whose /m/ sound is spelled *m* beside the **"m"** headings, words whose /m/ sound is spelled *me* beside the **"me"** heading, words whose /m/ sound is spelled *mb* beside the **"mb"** heading, and words with the *sm* consonant blend beside the **"sm"** heading.

m ______________________________

m ______________________________

me ______________________________

mb ______________________________

sm ______________________________

Spelling Words (/m/ and the *sm* blend)

G	E	M	O	G	C	L	I	M	B	U	O
U	K	L	I	S	K	S	U	T	F	B	X
F	U	O	D	M	T	O	A	L	A	R	M
I	T	F	M	E	T	Y	V	D	X	H	Z
R	P	L	C	L	D	A	M	A	G	E	A
E	I	A	F	L	D	O	A	L	M	N	F
M	A	M	Q	V	H	E	L	M	E	T	I
A	F	E	Z	Q	F	H	P	K	G	T	R
N	S	L	W	E	M	A	T	C	H	W	E
X	M	A	S	K	M	E	P	H	I	Q	M
I	R	Q	E	I	P	X	Y	V	F	M	E
S	M	O	K	E	Z	H	O	M	E	J	N

Fire Safety Words

1. fireman
2. firemen
3. mask
4. match
5. helmet
6. alarm
7. damage
8. home
9. flame
10. climb
11. smell
12. smoke

Weekly Oral Language

1. Correct the asking sentence.

   what is that noise

2. Correct the telling sentence.

   the fire alarm is sounding

3. What is the subject of the telling sentence?

   ____

4. Read the *telling sentence* again and identify the helping verb and the verb (action word).

   Helping Verb: ______________________

   Verb: ______________________________

5. Write your own sentence about this week's theme.

   ____

Instruction: *Teaching Children about Capitalization, Punctuation, Apostrophe s, and Contractions (Language Book 2)* by Habakkuk Educational Materials can be purchased separately to give students the knowledge they need to complete this worksheet. After students have completed the worksheet and turned it in, complete a copy of it on your classroom screen.

Parts of Speech

Directions: Identify what parts of speech are used in the sentence below and write them in the spaces provided. Then identify the subject (naming part) and predicate (action part) of the sentence and write them in the spaces provided.

Parts of speech: Noun marker, noun, pronoun, verb, helping verb, adjective, adverb, preposition, conjunction, interjection

Fire	**needs**	**oxygen**	**to**	**burn.**
__________	__________	__________	__________	__________

Subject (Naming Part)

____

Predicate (Action Part)

____

Instruction: *Teaching Children about Parts of Speech (Language Book 3)* by Habakkuk Educational Materials can be purchased separately to help students quickly learn the parts of speech. After students have completed this worksheet and turned it in, complete a copy of it on your classroom screen.

Name: ____________________

Addition and Subtraction Facts

Directions
1. Count the firetrucks above the first space and write the addend.
2. Count the firetrucks above the second space and write the addend.
3. Count all the firetrucks and write the sum after the equal sign.

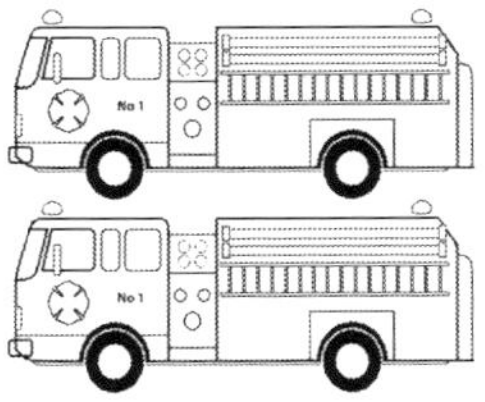

_______ + _______ = _______

Directions
1. Count the firetrucks above the first space and write the number.
2. Cross out 0 firetrucks.
3. Count the number of firetrucks left and write the answer after the equal sign.

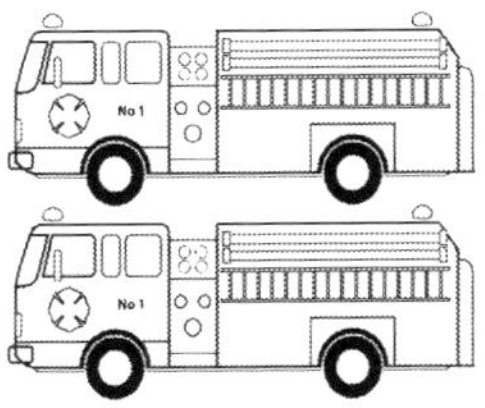

_______ − 0 = _______

Note: These cards can be used for the spelling word games listed on pages 21-22.

fireman	firemen
mask	match
helmet	alarm

Note: These cards can be used for the spelling word games listed on pages 21-22.

damage	home
flame	climb
smell	smoke

VOLUMES 1 AND 2

AAVAA. "Balloon icon set." Stock.adobe.com. Adobe Stock. https://stock.adobe.com/images/balloon-icon-set-party-balloon-icon-decoration-birthday-vector-balloon-black-silhouette-icon/440516985.

Alya. "Dove icon vector logo design template." Stock.adobe.com. Adobe Stock. https://stock.adobe.com/images/dove-icon-vector-logo-design-template/371102988.

Alyona. "Modern clothes pin vector icon." Stock.adobe.com. Adobe Stock. https://stock.adobe.com/images/set-clothes-peg-wooden-household-object-in-cartoon-style-isolated-on-white-background-modern-clothes-pin-vector-icon/479129215.

Anuwat. "Cute bee cartoon character vector Hexagon honeycomb and flower isolated on white background." Stock.adobe.com. Adobe Stock. https://stock.adobe.com/images/cute-bee-cartoon-character-vector-hexagon-honeycomb-and-flower-isolated-on-white-background/418949506.

Aroastock. "Illustration of baseball player playing." Stock.adobe.com. Adobe Stock. https://stock.adobe.com/images/illustration-of-baseball-player-playing/175419102.

Arturaliev. "Happy Presidents Day, Abraham Lincoln and George Washington." Stock.adobe.com. Adobe Stock. https://stock.adobe.com/images/happy-presidents-day-abraham-lincoln-and-george-washington/195710368.

Azuzl. "Long tail jellyfish object coloring book linear drawing isolated on white background." Stock.adobe.com. Adobe Stock. https://stock.adobe.com/images/long-tail-jellyfish-object-coloring-book-linear-drawing-isolated-on-white-background/428666427.

Bigmouse108. "Painter Sign Thin Line Icon Set." Stock.adobe.com. Adobe Stock. https://stock.adobe.com/images/painter-sign-thin-line-icon-set-vector/430467078.

Good Studio. "Cute male volunteer feeding stray dog in animal shelter or pound." Stock.adobe.com. Adobe Stock. https://stock.adobe.com/images/cute-male-volunteer-feeding-stray-dog-in-animal-shelter-or-pound/286093760.

Good Studio. "Peaceful landscape with trees, sun, cloud and birds." Stock.adobe.com. Adobe Stock. https://stock.adobe.com/images/peaceful-landscape-with-trees-sun-cloud-and-birds/278709697.

Good Studio. "Smiling bride and groom drawn with contour lines on white background." Stock.adobe.com. Adobe Stock. https://stock.adobe.com/images/smiling-bride-and-groom-drawn-with-contour-lines-on-white-background-portrait-of-happy-newlyweds-standing-together-romantic-wedding-celebration-monochrome-vector-illustration-in-linear-style/269996392.

Gstudio. "World book day." Stock.adobe.com. Adobe Stock. https://stock.adobe.com/images/world-book-day/262932143.

Hadeev. "Zoo Jungle, safari animals coloring book educational illustration for children." Stock.adobe.com. Adobe Stock. https://stock.adobe.com/images/zoo-jungle-safari-animals-coloring-book-edicational-illustration-for-children-set-cute-lion-crocodile-monkey-elephant-camel-rhinoceros-gorilla-hyppo-vector-white-black-cartoon-outline/434096184.

IMG Visuals Icons. "Sledding color icon." Stock.adobe.com. Adobe Stock. https://stock.adobe.com/images/sledding-color-icon/299432854.

IMG Visuals Icons. "The Flood Bible Story Icon." Stock.adobe.com. Adobe Stock. https://stock.adobe.com/images/the-flood-bible-story-icon-noah-ark-sacred-ship-in-worldwide-water-religious-legend-holy-book-scene-plot-biblical-narrative-glyph-chalk-linear-and-color-styles-isolated-vector-illustrations/310704435.

Kristina. "Cute kawaii cartoon tooth with toothbrush and toothpaste." Stock.adobe.com. Adobe Stock. https://stock.adobe.com/images/cute-kawaii-cartoon-tooth-with-toothbrush-and-toothpaste-coloring-page-for-kids-about-dental-hygiene/268289288.

Matthew. "Cartoon rooster." Stock.adobe.com. Adobe Stock. https://stock.adobe.com/images/cartoon-rooster/548769052.

Matthew. "Outline of a pot of stew." Stock.adobe.com. Adobe Stock. https://stock.adobe.com/images/outline-of-a-pot-of-stew/548755823.

Mhatzapa. "Set of sea animal kawaii doodles." Stock.adobe.com. Adobe Stock. https://stock.adobe.com/images/set-of-sea-animal-kawaii-doodles/370884066.

MissTiina. "Fire Safety." Stock.adobe.com. Adobe Stock. https://stock.adobe.com/images/fire-safety-set-of-illustrations-with-fire-truck-kids-dog/540768542.

Muhammad. "Vector Umbrella Outline Icon Design." Stock.adobe.com. Adobe Stock. https://stock.adobe.com/images/vector-umbrella-outline-icon-design/435400334.

N_A_Z. "Cleaning Service Lady Character Mascot." Stock.adobe.com. Adobe Stock. https://stock.adobe.com/images/cleaning-service-lady-character-mascot/349789963.

Nut Creator. "Cute design animal vector set 1 (Outline)." Stock.adobe.com. Adobe Stock. https://stock.adobe.com/images/cute-design-animal-vector-set-1-outline/436018825.

Pambudi. "Store vector icon in trendy flat style." Stock.adobe.com. Adobe Stock. https://stock.adobe.com/images/store-vector-icon-in-trendy-flat-style/337402148.

Panasovsky, Oleksandr. "Scallop seashell." Stock.adobe.com. Adobe Stock. https://stock.adobe.com/images/scallop-seashell/484863694.

Panasovsky, Oleksandr. "The human ear." Stock.adobe.com. Adobe Stock. https://stock.adobe.com/images/the-human-ear/457585494.

Roossoo. "Crack liberty bell Philadelphia icon symbol logo design vector line art style." Stock.adobe.com. Adobe Stock. https://stock.adobe.com/images/crack-liberty-bell-philadelphia-icon-symbol-logo-design-vector-line-art-style/426896893.

Sashimi. "A happy family of four, upper body." Stock.adobe.com. Adobe Stock. https://stock.adobe.com/images/4/242223728.

Ssstocker, "Cartoon red ringing alarm clock." Stock.adobe.com. Adobe Stock. https://stock.adobe.com/images/cartoon-red-ringing-alarm-clock/198157657.

Studio MRPOR. "Sun Face Sunglasses Summer Travel Beach Sea Planet Line." Stock.adobe.com. Adobe Stock. https://stock.adobe.com/images/sun-face-sunglasses-summer-travel-beach-sea-planet-line/585509842.

Svitlana. "Black-and-white illustration, set of corn stalk and corn cob with individual grains." Stock.adobe.com. Adobe Stock. https://stock.adobe.com/images/black-and-white-illustration-set-of-corn-stalk-and-corn-cob-with-individual-grains-hand-drawn-obje-t/401262258.

Tukinoto. "A woman in a suit giving a prohibition order." Stock.adobe.com. Adobe Stock. 禁止の指示をするスーツの女性 Stock Vector | Adobe Stock.

Twins Design Studio. "Coronavirus hand drawn icons, Cute health care doodle collection." Stock.adobe.com. Adobe Stock. https://stock.adobe.com/images/coronavirus-hand-drawn-icons-cute-health-care-doodle-collection-health-care-icon-set-corona-virus-disease-covid-19-infographic-design-element-vector-illustration/331862632.

Vacca, Alice. "Cute cartoon black and white thanksgiving day roasted turkey vector illustration for coloring art." Stock.adobe.com. Adobe Stock. https://stock.adobe.com/images/cute-cartoon-black-and-white-thanksgiving-day-roasted-turkey-vector-illustration-for-coloring-art/266945937.

Valenty. "Horse and rider on horseback logo." Stock.adobe.com. Adobe Stock. https://stock.adobe.com/images/horse-and-rider-on-horseback-logo-continuous-one-line-drawing/426397749.

Vectorfarmer. "Eagle Mascot Logo Spread Wings Vector Illustration." Stock.adobe.com. Adobe Stock. https://stock.adobe.com/images/eagle-mascot-logo-spread-wings-vector-illustration/244970879.

Vector Tradition. "Vector Saint Patrick with shamrock clover leaf." Stock.adobe.com. Adobe Stock. https://stock.adobe.com/images/happy-patricks-day-greeting-and-party-beer-pub-icons-vector-saint-patrick-with-shamrock-clover-leaf-leprechaun-in-hat-with-beard-and-mustache-coins-cauldron-and-horseshoe-with-stars-drum-and-hat/318913305.

Victoruler. "Milky Way." Stock.adobe.com. Adobe Stock. https://stock.adobe.com/images/milky-way/538897139.

Ville. "won" image on page 201 of Volume 2.

Zakowski, Igor. "Cartoon numbers set coloring book page with fruits." Stock.adobe.com. Adobe Stock. https://stock.adobe.com/images/cartoon-numbers-set-coloring-book-page-with-fruits/317607895.

Zakowski, Igor. "Cartoon safari animals coloring page." Stock.adobe.com. Adobe Stock. https://stock.adobe.com/images/cartoon-safari-animals-coloring-page/162239143.

Дарья Рымарь. "Set of simple five human feelings." Stock.adobe.com. Adobe Stock. https://stock.adobe.com/images/set-of-simple-five-human-feelings-five-senses-body-parts-mouth-or-lips-hand-and-nose-eye-and-ear-icons-sight-and-smell-touch-hearing-and-taste-senses-n-stock-vector/403262570.

Ольга Мещерякова. "Sense organs flat vector icons." Stock.adobe.com. Adobe Stock. https://stock.adobe.com/images/sense-organs-flat-vector-icons/220639410.

Made in United States
Orlando, FL
06 November 2026

90301619R00262